RETURN FROM EDEN

by George Moakley

Paperclip Publishing, LLC
Chandler, Arizona

Return from Eden

Published by: Paperclip Publishing LLC

Editor: Abigail T. Matteson
Cover Illustration: Shaun Cochran
Cover Design and Interior Typography: Hannah Thigpen

Library of Congress Control Number: 2026932614

ISBN: 979-8-9987821-8-3 (paperback)
ISBN: 979-8-9987821-7-6 (hardcover)
ISBN: 979-8-9987821-9-0 (eBook)

Printed in Rephen Printing, Co. LTD in Guangzhou and the United States of America

First Printing: 2026

Paperclip Publishing LLC
3800 W Ray Road Suite 5
Chandler, AZ 85226

www.paperclippublishing.com

To Diana, for believing in me.
To my kids, for sharing my love of monster stories.
To my readers, for requesting this story.

CONTENTS

FORWARD

If you've read and (hopefully) enjoyed *Kraken of Eden* and/or *Tides of Eden*, you may confidently skip this forward and begin reading *Return from Eden*. But if you're new to the *Deep Space Exodus* series, or if you want a bit more insight into the foundation of the series (or maybe you're just a compulsive reader like me), then please continue.

I originally envisioned *Kraken of Eden* as a standalone novel, wondering, at first, whether I would really be able to produce a full-length book. When I finished writing, I had over 120,000 words and a number of additional plot ideas that would have made the novel prohibitively long. But after publishing, I found myself missing characters I'd grown fond of, wondering about their future. As reviews were posted, it became clear readers felt the same way.

So, what to do? Keep writing, obviously!

As I began storyboarding *Tides of Eden*, I thought of a three novel story arc. I'm resistant to think in terms of sequels or trilogies, largely because I want any reader to be able to pick up any *Deep Space Exodus* novel without needing to read its predecessors. I've also, as a reader, found the second novel often feels more like a "bridge" than a complete story. While I hope readers of *Tides of Eden* will want to read *Kraken of Eden*, my aspiration was a story that worked on its own.

While writing *Tides of Eden*, I continued to do book events for *Kraken of Eden*, and I met more and more people who loved *Kraken of Eden* and wanted to share their great ideas for additional novels set in the same future timeline. Now, one might ask, "how many stories could there possibly be," to which I might respond, "how many monster stories have been set on Earth?"

Return from Eden was requested by quite a few *Kraken of Eden* readers, and I hope you enjoy reading it as much as I enjoyed writing it. While I generally retain my commitment that each novel will be a complete story and can be read in any order, *Return from Eden* has a special relationship with *Kraken of Eden*. It could be read as an independent novel, but I believe readers would enjoy the novel more if *Return from Eden* is read after *Kraken of Eden* and *Tides of Eden*.

Though the novels work as standalone stories, there is a grander arc to the series. Of the fan suggested additional plots, three have been selected, including this one, based on their fit within that bigger story. So, as of this writing, this is the third of six planned novels.

Unfortunately, this creates a challenge. In my view, great stories take the reader to an alternate reality. Whether science fiction, historical fiction, organized crime stories, etc., immersion in a rich, plausible world unlike one's day-to-day existence is a foundational quality of the experience. But such stories involve world building. How do you revisit this alternate reality without either boring returning readers or bewildering new ones?

One option is to write a brief forward.

I'm a near-term pessimist and a long-term optimist. I believe that, over the next hundred years or so, humanity will be severely tested by climate change, famine, drought, pandemics, mass extinctions, and other ramifications of global population growth. I also believe that our descendants, confronted by these challenges, will finally realize if we don't reduce Earth's population to a sustainable level, Earth will do it for us in very unpleasant ways. I believe that these challenges will, finally, drive humanity to colonize our moon, Mars, and, eventually, nearby stars. This is why I refer to these novels as *Deep Space Exodus* stories.

I enjoy hard science fiction. Plausibility adds so much to the story, especially for a horror story. So, as much as I love space operas involving hyperdrives and warp speeds, I wanted to explore a future without faster than light travel. I gave a great deal of thought to the implications of humanity aspiring to spread among the stars without faster than light travel, and I found myself thinking about humanity spreading through Oceania. If people could spread one land mass at a time, through islands, archipelagoes, and eventually the Australian continent, then could we not, someday, spread one star system to another? If, hundreds of years ago, when we came of age in our teens and typically reached senescence in our forties or fifties, some of us were willing to spend two to three years of their lives crossing oceans by wind, would we not, with significantly longer lifespans, be willing to spend ten to fifteen years crossing interstellar space? And if those crossings were at relativistic speeds, how would time dilation affect our culture?

These novels take place in the 26th century. Earth's leading nations long before formed a Deep Space Cooperative to share the costs and profits of interstellar exploration and colonization, including a program to preclude tribalism by monitoring colonial gene pools and incentivizing migration between worlds. Earth's ecosystems start to recover as migration reduces her global population to a sustainable level, but given the mass extinctions we're currently driving, these ecosystems are, sadly, different from the ones that nurtured our ancestors. Some two billion humans now live elsewhere in our solar system, and seventy-five million more live among colonies across dozens of star systems.

For me, though, the most important thing was xenobiology. What would it be like to explore a truly alien world? What could be different about alien life? What could be the same? Humans haven't discovered alien life…yet. Many experiments have recreated primordial conditions and demonstrated that the basic building blocks of carbon-based life form naturally. From this, I believe it is reasonable to extrapolate that, given sufficient time and the right conditions, such organic compounds would combine to form simple cells comparable to earthly bacteria that would in turn evolve to more complex cells and multicellular life. There may be somewhere in our universe "life" based on some other chemistry, but surely there must be an abundance of alien life based on the same chemistry as ours. But, even within the boundaries of "the same chemistry as ours," there can be tremendous diversity. It seems to me that life on Earth is remarkably consistent biochemically, using the same nucleic acids and the same amino acids, even the same stereoisomers of complex chemicals. Surely, just as there must be an abundance of alien life based on the same chemistry as ours, the odds against such life using the same specific chemicals would be astronomically high (pun intended)!

So, in our stories, as humanity spreads through and settles nearby star systems, we encounter many worlds with organic chemicals of scientific and commercial interest as well as a smaller number that have simple cells. But, in our stories, Eden is the first (and only) alien world discovered with a complex ecology. Blue skies, blue oceans, green leaves, creating the opportunity for delightful thought experiments about what truly alien life might be like. Bear in mind, as you read about our colonists discovering the denizens of their new home and our xenobiologists cite earthly analogs, that all of the earthly creatures offered as examples are real!

I hope you enjoy reading *Return from Eden* as much as I've enjoyed writing it, and I hope you're intrigued enough to read *Kraken of Eden, Tides of* Eden, and all of the forthcoming novels set in this future timeline.

Thank you!

PROLOGUE

MESSAGE RECEIVED . . .

The harbor master rubbed her temples and muttered obscenities again.

"Fucking idiots are going to run into each other! It's just a matter of time!"

Her staff was used to her muttering. For her, muttering correlated with stress, and stress correlated with colonial growth. Most of them weren't close enough to make out exactly what she was muttering, but they didn't need to. Many of them were muttering, too…

Alexandria, like any colony, generally celebrated growth. Just a few days ago, their population surpassed 200,000 with the birth of a child appropriately named "Alexander." But growth meant more spaceport traffic. A second orbiting platform was under construction to provide an additional space port, but not nearly quickly enough. And she just learned they were expecting *another* deep space vessel!

She snorted and shook her head. *Maybe a spectacular collision would speed up construction!*

She glanced about the room. Nobody was looking, but she had the feeling they were making a point of not looking. She sighed, closed her eyes, and pinched her nose to collect herself.

Another seventy-two minutes. Just seventy-two minutes until the end of my shift.

She indulged, briefly, in a bit of daydreaming about her plans once she finished her shift. She'd start by sitting quietly in her quarters, lights dimmed, with an adult beverage. Then, she'd invite her favorite to join her. The harbor master shook her head, squared her shoulders, and resumed pacing around the massive spherical projection depicting the spaceport, pausing here and there to squint occasionally, searching for problems to address before they got out of hand.

At the center of the primary projection was the rotating drum of the colonial spaceport. Her control room occupied part of the command ring, the ring facing the planet surface. From the center of the command ring, the space elevator tethered the drum to the primary dome of the sprawling surface settlement. She turned from the projection to look through observation windows. Dusk crept across the settlement, and the domes glowed blue through the perpetually stormy primordial atmosphere. Lightning bolts flashed among the clouds, competing with the flashing lights of the space elevator against the blackness of space.

The control room stretched across the width of the command ring and extended far enough in either direction to see the floor visibly rise with the curvature of the drum. Surrounding the projection, her staff stood at their workstations, dividing their attention between miniature projections highlighting their respective areas of responsibility and the primary, about which she resumed pacing.

Walking slowly around the projection, she could see hanger bays in the uppermost ring of the rotating drum opening and closing for shuttle tugs and landing craft. She could see space buoys surrounding the drum, some idle, the rest hitched to a broad variety of craft. She could see pressurized transit hubs facilitating the zero-g movement of people, supplies, and cargo between vessels docked to them. Surrounding all of *that* was a cacophony of shuttle tugs, landing craft, interplanetary craft, and private yachts. The shuttle tugs, landing craft, and interplanetary craft didn't stress her. Their pilots were professionals who knew their stuff, followed the rules, treated each other courteously, and knew the penalties for violating port code of conduct. It was the private yachts that stressed her. The colony was growing rapidly, and with it, the number of wealthy morons piloting pleasure cruisers. Status symbols for those with the means, which meant they, all too often, bought more "ship" than they could handle, especially with a minimally passing score on their pilot's exam. Morons that didn't quite grasp that the port code of conduct applied to them and wasn't just a set of suggestions.

These morons accounted for her baseline stress. A growing source of stress, to be sure. But the current escalation in her muttering, the recent complication compounding her stress, was the deep space vessel *Eureka*.

Interstellar craft were a navigational hazard for any spaceport. For one thing, they're immense. For another, while in port, they swarmed with shuttle tugs. *Eureka* was a standard configuration. A drum the size of the colonial spaceport rotated about a 2,500 meter shaft. That shaft began and ended in massive funnels to collect interstellar matter, then annihilate it with anti-matter for propulsion. Three weeks ago, the *Eureka*, having decelerated to cruising speed, was greeted by shuttle tugs dispatched to tow her to a space buoy as far as possible from the orbiting platform.

Once the shuttle tugs tethering the *Eureka* to her buoy ferried her 200 migrants from the Panmictic Incentive Program (PIP) and crew of fifty to the orbiting platform, she was surrounded by many more shuttle tugs. Some were refitting her, bringing her up to date with the last fifteen years' technological advancements. The rest were reconfiguring her drum. Her next crossing would be a new colony start, so the tugs were replacing her rings of cargo pods with residential, commercial, industrial, and agricultural rings.

Despite being at the outer edge of the space port, the deep space vessel was an attractive nuisance for rubbernecking morons. Refits and reconfigs were delicate operations testing the skills of even the most experienced shuttle tug operators, especially when those operations involved massive ring segments and individual space workers in environmental suits. The last thing the shuttle tugs needed was the added stress of barely competent pleasure craft pilots zooming by, too close and too fast.

The harbor master had hoped to see the *Eureka* on her way before another deep space vessel approached, but she was behind schedule. And now, it looked like the *Lucky Strike*, returning from Eden, would arrive before the *Eureka* departed. Setting up a second deep space vessel compatible buoy in time for the *Lucky Strike's* arrival was a challenge and yet another attractive nuisance.

Damn!

The harbor master paused her pacing as she approached the *Eureka* and noticed a flashing red icon. She swore and spread her hands to zoom in on the icon. It gradually resolved until she could see it was a pleasure cruiser moving through the projection towards the *Eureka*. She highlighted the cruiser and called out, "who's talking to the idiot piloting that yacht?"

After a moment, a voice responded, "I am!"

"Well, tell that moron he's too close to that deep space vessel! He's going to run right into the shuttle tugs!"

"It's a 'she', not a 'he', Harbor Master."

She turned, stared, paused to collect herself, then quietly, but stridently, said "I don't give a rat's ass whether it's a 'he' or a 'she' or a fucking puppy! Tell that dunce she's too close to that deep space vessel, *right now*!"

"Yes ma'am!"

As she watched, the pleasure cruiser slowed, then veered away. She sighed, not realizing she'd been holding her breath. Less than an hour left in her shift!

"Harbor Master?"

She turned to the voice and saw a young man looking up from his comms station.

"Yes? What is it?"

"That inbound deep space vessel? The *Lucky Strike*? It must have slowed enough for transmission. I'm receiving a distress signal!"

She stared at him in disbelief. In all the years she'd been harbor master, she'd never received a distress signal from a deep space vessel. She'd never even *heard* a distress signal from a deep space vessel! Her irritation fell away as professionalism took hold. They had protocols for anything and everything, even the things that never happened.

"You say she's decelerating?"

"Yes ma'am. Appears to be on a normal approach."

She nodded. The ship would still be a few days away. Plenty of time to process whatever was going on, and there wasn't much she could do until the ship was closer. For now, the ship's deceleration enabled message exchange, but it was still too far away for interactive communication.

"Is there a message accompanying the distress signal?"

"Yes ma'am."

She nodded. The message should have the information they'd need to prepare.

"Route the message to me. Get my relief up here ASAP. I'll alert the Governor."

PART 1
LUCKY STRIKE

DAY 1

PROMOTION

Sri's father always said, "Even the greatest transitions are still frightfully boring while you're living through them." Sri wondered whether anyone would remember his father now. In his day, nobody thought him boring. His father had been an intermittently successful entrepreneur, with a lifetime of making and losing fortunes. Something of a minor celebrity on their home world. Not a high bar to hit back then. In Sri's youth, it had been a small colony. Now, centuries later, the population was in the millions. He often wondered whether his father was right about great transitions being boring, or whether, for someone like his father, the novelty of *novelty* wore off.

He missed his father.

Sri understood, intellectually, that launching a career as a deep space officer meant leaving everyone and everything he'd ever known behind. Interstellar crossings, even at relativistic speeds, took ten to fifteen years. Sure, as they approached the speed of light, time slowed aboard ship, but not for friends and families. It hit him harder than he'd expected, emotionally, to see everyone from his youth age so much during each crossing. When he inevitably learned of his parents' passing during one of his crossings, it crushed him.

Of course, deep space careers had their rewards. They paid well, especially as his career progressed to senior ranks; investments compounded nicely during his crossings. Such rewards enabled many deep space lifers to retire young, or take long breaks from their careers to raise families. But, eventually, as their children left the nest, they often returned to their careers and lost track of their descendants, just as he'd lost track of his parents, siblings, and their descendants.

But not Sri. It wasn't a choice to not have a family. It just...never happened. Just one regret amidst the many regrets of an otherwise amazing life among the stars.

Sri sighed. With aging comes perspective, and, all too often, regret for the chances not taken, the choices not made. Odds weighed against one's risk tolerances, decisions made not in the pursuit of dreams but the fear of failure. Decisions that, sometimes, do not hold up well in the illumination of hindsight. But, time heals all wounds, especially when coupled with burning ambition. And, sometime last night, his greatest ambition was finally attained.

Captain Sridharan Rizzo.

Sri shook his head. He was walking towards the captain's conference room.

My conference room, now!

He'd known the promotion was coming, of course. But, he didn't know *when* it would come through. If he still had any close relatives, he could forward the announcement to them; but, after all these years, all these crossings, there were no relatives close enough to care.

So, this morning, when Sri woke and saw the announcement, he had a cascade of feelings. Joy, pride, satisfaction, relief, loneliness, disappointment, and a dash of depression. But, with a smile and a nod, he dressed, broke his fast, and prepared himself for his day just as he'd done every morning for, well, a long time. A very, *very* long time...

As he prepared for the very auspicious day ahead of him, Sri decided to head to his conference room a little early so he could savor the experience. Sadly, as his father would have observed, the long walk towards adventure was rather dull, and lonely, given how empty the corridors were at this hour. He entered the room, and, as the lights automatically came on, startled one of the ship's cats.

Must be earlier than I'd realized!

Sri stamped his foot to scare the grizzled orange tabby out the portal into the corridor. He glanced about, half expecting to see whatever nasty things the cat was presumably hunting. Thankfully, he saw nothing. He found typical shipboard vermin disgusting. Relieved, he walked to the coffee dispenser and reached for his mug.

It was truly *his* mug, custom made to fit his hand. An ungainly thing made years ago by his shipmates to commemorate his unrequited love for the sea, its base resembled surf, its sides were sky blue with wisps of clouds, and its handle was a sextant. Sri's love of the sea was as inexplicable as it was unrequited. He'd never been to Earth, never laid eyes on her seas. He was born on a lifeless rock among the inner colonies more than two centuries ago and recently celebrated his 94th birthday.

Since boyhood, he dreamt of a career crossing vast oceans of interstellar space at relativistic speeds and developed a fascination for Earth's great explorers. He saw parallels between their aspirations and his own, an empathy for those that committed years of their lives to transoceanic voyages driven by wind. And, now, he too was a captain! Captain of a great ship making great crossings through the black, discovering new worlds.

Finally!

Sri set down his mug as he relished the thought. It was a rank rarely reached, a rank achieved by precious few. With a tilt of his head, he silently acknowledged that, on the other hand, those that *did* make captain usually did so at a younger age…

Sri shook his head. *Not a productive line of thinking!* He leaned forward and filled his mug. He sipped, walked slowly around the captain's conference table to the captain's chair at the head of the table, and took his seat. Senior staff, *his* senior staff, would be arriving soon for their final meeting before departure.

Years before, he'd sat in the seat now to his right: the Executive Officer's chair. They were approaching Alexandria, with 200 PIP migrants, a crew of fifty, and a wealth of cargo. A typical crossing of nearly three years for those aboard and a bit more than fourteen for everyone else. As Captain Lubandi entered the room, they'd shared plans to explore Alexandria, to spend a year or so burning through their bonuses, getting into a modest amount of trouble before signing on to another outbound deep space vessel.

Then, everything changed.

When they decelerated enough to receive messages, Captain Lubandi played what they presumed would be a traditional, routine greeting from the spaceport's governor. Instead, they learned that, after centuries of exploration, a world had been found with a complex ecosystem. The first, in all of known space, with a rich, complex biosphere.

Blue skies. Green leaves. And blue seas! The planet Eden.

The Deep Space Service (or DSS) was anxious to start a colony there, so the crew of the *Lucky Strike* was asked to stay together as a team, and, after an expedited refit, take the first 1,000 colonists to Eden. Sri remembered the moment well. Outwardly, he remained stoic. Inwardly, his mind raced as two absolutely amazing scenarios played out in his mind.

First, this was an initial colony run! They'd expand the ship's complement to 100 to babysit the colonists and all their paraphernalia. Seventy-five would stay at the new colony to form the government, leaving twenty-five for the return journey. The ship would be refitted, exchanging the rings supporting the colonists for rings of cargo harvested by the autonomous drones that initially discovered Eden. Return trips from new colony starts were rare opportunities for advancement. Pure cargo runs were usually flown autonomously, their routes optimized without concern for the frailties of a human crew. But deep space vessels had to return from new colony starts and they must be crewed, so return trips from new colony starts were golden opportunities for promotions. Captain Lubandi would stay with the colony to serve as interim governor, and Sri, her Executive Officer, would be promoted to captain. Captain Rizzo, in turn, and with consultation from his former peers, chose twenty-four souls to accompany him, promoting a number of them to senior staff positions.

Second, this was a world with oceans! As exciting as it was to finally earn his captaincy, Sri was equally excited about this unprecedented opportunity to actually *see* oceans! It was the thrill of a lifetime!

As soon as they arrived, xenobiologists and planetologists boarded landing craft to explore the surface. Sri, knowing the *Lucky Strike* carried watercraft, began his campaign. He convinced the pilot of the first oceanic expedition to bring him along and even, albeit briefly, take the helm. Cruising near a coastline between scattered islands clad in green, they stopped to let the science team explore an interesting pinnacle. Sri and the pilot stood watch as they submerged; Sri's mind reeled as he looked out across endless water stretching to the horizon. All too soon, one of the patrol drones circling the divers reported the approach of a large marine creature. As the divers returned, they prepared to move on to another site. But the sea creature attacked the boat! Nobody was hurt, but their landing craft had to rescue the damaged boat. Thus ended Sri's first (and almost certainly only) marine command on a rather ignominious note.

Sri's musings were interrupted by the arrival of Amar Aggrawal, who blushed and said, "I know; I know. I made an extra effort. I didn't want to be late for *this* meeting!"

Amar, their astrogator, was habitually tardy and unkempt. Most presumed it had to do with his time spent in the deep abstractions of plotting interstellar courses and let it go. But, for some reason, it always irritated Captain Lubandi. She hid it reasonably well; but, after years serving together, the senior officers could tell; it became a source of some amusement for everyone, except Elke and Amar.

As Amar took to his seat, Sri pointed to his mug, then the coffee dispenser with a raised eyebrow. Amar shook his head and resumed tapping at his wrist computer. Sri briefly considered the daunting effort of attempting small talk with Amar, but he was spared by the arrival of Emil Nguyen. Emil had been one of Nikko's Operations leads, and was now Sri's Executive Officer. Emil was followed by Chief Engineer Katya Smith and Chief Medical Officer Simon Young. Without colonists or migrants, there was no need for an operations team or a security officer.

As Sri's Executive Officer, Emil was second in command, and once underway, would stagger his sleep cycle so he could be on duty while Sri slept. Not that there would be much to do, of course, not during a cargo run. He was young, in his fifties, tall, athletic, and physically imposing, with brown eyes, brown skin, and short cropped red hair. Despite Sri's best efforts, Emil remained aloof. He'd tried various ways to break the ice, to build a bond, but to no avail.

Well, there's plenty of time for that!

Simon, blonde, olive skinned, quiet, and a bit overweight, was also young. He was not Sri's first or second choice for Chief Medical Officer, but most of John Schmitt's medical staff wanted to stay with the new colony. He was a fully fledged doctor, but he'd never held a supervisory position. But one of his two physician's assistants, Yvette, was very experienced and could help Simon learn along the way.

Sri smiled thinking of Yvette.

Sri was also intrigued by Katya. She was a bit older but still in her eighties, relatively short and a bit pudgy, with fair skin, high cheekbones, and short blond hair. She was a very talented engineer who Moira Tam rewarded with promotions along the technical track. But Moira's engineering managers all wanted to stay with Eden colony. Speculation was they thought it only a matter of time before Moira succumbed to the itch and signed on for another crossing, leaving one of them to take her place. For the return trip, Katya, the most senior individual contributor, was offered promotion to Chief Engineer. To Moira's surprise, she accepted eagerly.

Sri knew this would be a difficult transition for anyone. Katya wasn't the only one who would need a lot of coaching. Three engineering team leads, also new to leadership positions, each managed five engineers, who now reported to Katya. Blind leading the blind, but it couldn't be helped.

They'd been running pre-flight simulations for a week now, and, to Sri's relief, Katya's engineers performed excellently. And, of course, barring unforeseen and highly unlikely complications, there really wouldn't be that much to do except reconfigurations. In port, the ship's drum rotated to simulate gravity, so "down" was "outward." They'd spend the first half of their journey relentlessly accelerating at 1.2g, approaching but never achieving, the speed of light. The drum would not be rotating, because "down" would be aft. To prepare, Katya's engineering teams would reconfigure the drum accordingly. They'd repeat the chore at the midpoint of their crossing, as the ship decelerated at 1.2g so that "down" would be forward, and then once again at arrival when the drum would spin again. Reconfiguring "down" was always a challenging chore. Given it was a cargo run, only two rings would need to be reconfigured; but, with inexperienced supervision, it would be interesting, to say the least!

Sri cleared his throat and addressed his new staff. "As you know, Governor Lubandi will be joining us shortly. Anything to report?" Smalltalk ceased as he looked about the room.

Emil shrugged. "No, Captain, everything's nominal. We're good to go." He paused, then a smile crossed his face. "Hey, it's official now, isn't it? Isn't it! Congratulations, *Captain* Rizzo!"

Sri stood, smiling awkwardly as the others congratulated him. He raised his hands, nodded, and thanked them. Deciding it was time to return to ship's business, he was about to ask a question when Governor Lubandi, renowned for her punctuality, arrived exactly on time. An impressive stature, she ducked to enter what had been, for three crossings and now nearly a decade, *her* conference room. Everyone stood as Sri smiled and greeted her.

"Governor Lubandi, thank you for gracing our final staff meeting before departure."

She smiled and waved them to their seats, taking her place to the right of the captain's chair. Sri turned to Amar and asked, "Amar, have you finished plotting our course?"

After a moment, Amar looked up. "Oh, yes. All plotted and entered. Quite similar, in fact, remarkably similar, to the outbound course. The relative movement of the two stars and the orbital positions of the two planets will result in a return total distance almost identical to the outbound journey. Assuming we leave on schedule, we'll arrive in fourteen years and, uh, sixty-eight days, which will be 764 days ship's time." He paused, then smiled awkwardly.

Elke turned to Sri and grasped his hand firmly. "Godspeed, Captain Rizzo."

He held her hand, then released it. “Thank you, Governor Lubandi.” After an awkward moment, Sri jabbed. “Seems anti-climactic, doesn’t it?”

Elke laughed lightly. “Trust me, I’ve done this a number of times. It always does.”

She turned to the rest of the table, and her voice filled the room. “It was an absolute pleasure flying with you. I’m sorry you can’t stay as we continue to explore Eden; but, we’ll keep you posted as you begin your acceleration until we can’t exchange messages, which I think will be in a week.”

The *Lucky Strike* officers came to attention and saluted her; she returned a salute in kind.

“Alright, we’re adjourned,” Sri resolved. “Governor, I’ll escort you to the hangar deck.”

As they walked through the hall, Sri whispered, “Thank you for not mentioning my boating misadventure.”

She laughed and said, “It didn’t seem like the time. And I doubt anything will be taking a bite out of the *Lucky Strike*.”

“One can never tell.”

The ship’s drum was only configured for the current complement, so it didn’t take long to reach the hangar deck. As they entered the hangar deck and approached Elke’s shuttle, he continued, “I expect our staff meetings will stay short; there’s only twenty-five of us.”

“Yes, that’s what I’d expect,” Elke nodded in reply. “And they’re good people. See if you can work on Amar’s discipline, will you?”

He laughed. “No promises. You knew, when he signed on, what he was like, didn’t you?”

She shrugged her shoulders. “He had an excellent reputation. And I thought I could make some progress with him.” She paused, then smirked. “He’s oblivious. Probably part of what makes him so good at his job.”

Sri laughed quietly in agreement. He smiled awkwardly, looking back and forth between the ship and Elke.

“Permission to hug the governor?”

Elke’s eyes went wide with surprise. She laughed and they briefly, awkwardly embraced.

“Thank you for this opportunity,” Sri said.

Elke shook her head. “You’ve earned it. You’ve a bright future ahead of you, Captain. I happen to know you’ve had some glowing performance reviews.”

"Well, that's good to hear. Well..." He paused, "I suppose it's time. I'll operate the hangar bay."

Elke nodded and boarded her shuttle tug. As the door closed, Sri went to the operator's room to oversee her launch and departure.

DAY 2

DEPARTURE

Sri, Emil, Katya, Amar, and Simon stood at their workstations surrounding the bridge display projection. The rest of the crew stood on the raised platform behind them. Katya's engineers were ready to leap into whatever action may be required of them, having rehearsed every conceivable scenario through endless simulations over the past week.

The *Lucky Strike*, tethered to a space buoy near the orbiting platform of the Eden colony, nearly filled the projection. The deep space vessel's rotating drum now consisted of a command ring and residential ring for the twenty-five souls taking her back to Alexandria. Thirty-eight rings of cargo storage and processing pods replaced the residential, agricultural, commercial, and industrial rings that, during the fifteen days since their arrival, were moved to the orbiting platform.

Beneath the *Lucky Strike* was the broad equatorial desert that would, one day, hold the colony's surface settlement. Surrounding the desert, to the north, south, and west, brown gradually gave way to green. To the east, massive, unnamed mountains created the forest by wringing dry nearly every cloud that tried to cross them from the ocean beyond, forming a lush, coastal rain forest.

Sri imagined the bustling spaceport Eden would someday become. The economic and scientific importance of Eden would drive faster than typical growth. Within a few years, a space elevator would link the orbiting platform to a surface base station surrounded by a rapidly growing city. Would there be domes, or would there finally, for the first time on a world other than Earth, be a city open to the sky?

He was suddenly consumed by regret. Yes, he'd finally earned his captaincy, but it was hard to leave so much potential behind. He'd have to find his way back, someday.

Sri shook his head and returned to the present. There was no surface settlement, not yet, nor a bustling space port. Just two shuttle tugs launching from the orbiting platform, preparing to tow the *Lucky Strike* to her departure point. Sri looked about the bridge. Every face was at least familiar. Some he knew from the crossing to Eden; some he'd gotten to know during the simulations. The rest he at least knew from reviewing their personnel files while finalizing the ship's roster. There was plenty of time to get to know them all, once they were underway. He returned his attention to the display.

Sri cleared his throat, then asked, "Emil, are we ready for departure?"

Emil looked at Katya, who nodded, then turned back to Sri. "Yes, Captain. We are ready!"

Sri smiled and nodded. "Very good. Emil, inform our escorts."

Emil tapped at his workstation briefly, then joined the others to watch the display. The shuttle tugs converged at the space buoy. Their robotic arms detached the lines tethering the *Lucky Strike* and attached them to their hulls. As they moved away from the colonial drum, the lines became taut, and the great ship began to move. Not that Sri could feel the movement. He nodded to himself, silently acknowledging the skill of the shuttle tug pilots.

He turned to Emil and asked, "estimated time to departure point?"

He consulted his workstation. "Sevety-three minutes," he responded.

Sri then turned to Katya. "Engines primed?"

"Aye."

"And you're ready to reconfigure?"

"Aye, Captain."

Sri nodded, then looked about the room and, with a wry grin, asked, "Anyone for a few hands of poker?" A few chuckled as the mood measurably lightened.

The minutes slowly ticked by as Eden fell away from the display. Katya leaned over to say something to Emil. Amar frowned as he tapped away at his console. Simon stepped away from his station, walked towards his physician's assistants, and started chatting. Other crew members began talking with their neighbors.

This is a good crew, Sri thought. He'd initially been concerned. Most colony starts weren't as enthralling as Eden. Usually, given the spartan conditions of a new colony, a return crew was selected from an overabundance of applicants. This time, they had to recruit. Assembling a crew required enhanced bonuses and less rigor than he would have liked with regard to compatibility modeling. But he'd been reassured by their performance during simulations.

When they were far enough from the orbiting platform, the shuttle tugs slowed, stopped, and detached their tow lines. They turned to make their way back to their orbiting platform bays. A disembodied voice from one of the shuttles wished them luck.

"Godspeed, *Lucky Strike*!"

Sri smiled and replied with a thank you. He turned to Emil and asked, "Are we clear?"

Emil, staring at his console, raised a finger. "Not quite yet…the shuttle tugs are returning to Eden…and…we are clear and ready for ignition!"

Sri looked around the bridge and cleared his throat, albeit a little awkwardly. He felt the need to say *something…*

"Centuries ago, the great explorers of antiquity looked across seemingly infinite seas. With a leap of faith and a hunger for discovery, they set forth across great oceanic voids in modest vessels constructed of timber and driven by wind in search of great wonders. Now, we too set forth across a great void, trusting each other and this fine ship to bring us safely to our destination!"

He turned to Katya and said, "Ignite the engines!"

"Aye, Captain!"

The ship shuddered lightly; on the display, the rear funnel began to glow.

"Cease rotation!"

"Aye, Captain!"

The drum's rotation gradually slowed as the ship accelerated, and their perception of "down" gradually shifted from the inner surface of the drum to the aft surface of each ring. The modular internal structures of the ship's working and living areas needed to be shifted accordingly.

"Emil, Katya, I expect a smooth reconfiguration!"

DAY 3

PERFORMANCE EVALUATION ..

"Katya, this was unacceptable."

Sri and Emil sat at the captain's conference room table with Katya. A section of the second ring was projected above the table.

"I know it was your first time supervising a reconfiguration. And I know it's a cargo run, so we don't have any migrants or colonials or, God help us, livestock squawking and complaining." Sri pointed at the display. "But this took far too long. And worse, this assembly came loose. Someone could have been hurt."

Katya nodded curtly, averting her eyes in embarrassment. She knew it had gone poorly. She'd entered the room expecting to be chewed out, but that wasn't Sri's style. He was calm, in coaching mode. Even so, this was embarrassing, almost unbearingly so, and undeserved! She wasn't the one that screwed up!

"I apologize, sir. That section was reconfigured by Morgan's team. One of her people didn't show."

Sri paused and furrowed his brow with brimming angst. The last thing he needed in a new crew was delinquents. "What do you mean, 'one of her people didn't show'?"

Katya raised her head. "One of her people, Robert Esposito, failed to report for duty. Morgan didn't notify me. She made the best of the situation but was short staffed and fell behind. Then she joined her team to get back on track, which constrained her ability to do adequate inspections. I wasn't informed of any of this until the assembly came loose."

She shook her head and leaned forward.

"I have already reprimanded her for not informing me immediately. I could have helped with inspections. Getting it right was more important than keeping to schedule."

Sri nodded slowly, then turned to Emil with a raised eyebrow. Emil shrugged. Sri turned back to Katya.

"Why didn't Esposito report for duty?"

"He wasn't feeling well. When he didn't report, she called him. She told him to rest, and, if he didn't feel better, to report to sick bay. Again, I didn't know about this at the time."

"And what have you learned?"

Katya pursed her lips, tapped the table a few times with an index finger, then folded her hands. She knew what Sri wanted to hear.

"I need to check on my people more often. I need to do spot checks."

"Those things would have helped. Did you emphasize getting things done quickly? Or getting things done right?"

Katya paused, then said, "Generally, I emphasize getting things done right. But…" she blushed briefly, then straightened in her chair. "But the engineering teams set up a competition. On their own! A bet, really. The slowest team would have to prepare breakfast for the fastest team for a week. The team leads came up with it, so I didn't find out about it until our prep meeting before the reconfig. I, uh, thought it would be motivating. We wanted the reconfig to go well. But, in hindsight, I can see where this would have prioritized schedule over quality."

Sri sighed and rubbed his temple. Emil snickered, but hid his enjoyment under his hand. Such childish behavior could cost them dearly in the future. Sri leaned forward.

"Katya, you're a very talented engineer. Among the best in the service. But this is a big promotion; you skipped being a direct supervisor. I can see now that there are things you would have learned as a direct supervisor that would have helped prepare you for this role." He could see, in her expression, that she was a bit offended.

"Please don't take this as criticism. If anything, I'm chastising myself for not realizing this. As I said, you're a very talented engineer. You're in this role because Chief Engineer Tam, Captain Lubandi, and I all believe you have the aptitude to excel here. But it *is* a big promotion, and that means you, as would anyone, have a lot to learn. Coaching you is my job, correct?"

Katya nodded. But it still stung. She shouldn't be here. She didn't want this responsibility. She understood how *things* worked, not how *people* worked! She didn't want this promotion. But she couldn't stay on Eden, and the only way to ensure a space aboard the *Lucky Strike* was to accept a promotion. She knew they needed a chief engineer.

No, not "accept" promotion. *Pursue* promotion. *Campaign* for promotion. She'd *fought* for a promotion she'd resisted for years, feigning a belated aspiration.

Damn you, Ravi!

She shook her head, realizing she needed to pay attention to Sri's lecture.

"There's more to leadership than holding your people accountable to goals," he rambled, "especially as you move up the ranks. Among other things, leaders establish a team culture, a conceptual framework within which people make decisions. Their understanding of relative priorities, the things they know you expect without having to tell them every time. This is absolutely essential in a crisis! That's when your people need to make the right decisions even when there's no time to consult with you or each other." He paused, eyebrows raised, until Katya nodded.

She could have stayed. She *should* have stayed. But there was no avoiding Ravi. That's the problem with the start of a new colony; there was no avoiding *anyone*.

Never should have gotten involved with him in the first place!

"The DSS has a leadership culture," Sri lectured. "Part of my role is to adapt that culture to the needs of this ship and crew. Part of your role is to do the same for your engineering teams. As Chief Engineer, you establish a culture for your direct supervisors. How they balance quality against performance, the needs of the team against talent development. Endorsing a performance wager is fine, as long as you're clear that quality remains everyone's top priority."

Sri waited for Katya to nod again.

"Good. Now, I agree with your assessment of Morgan's performance. I like that you took responsibility for her decisions, and I think you coached her well." He sighed, stood, and added, "So, how *is* Esposito?"

Katya and Emil stood as well. "I'm not concerned. Morgan said something about him feeling weak. I mean, he's always been a little pudgy; but, when we came aboard, you could see he was much trimmer. I guess he's been spending a lot of time in the gym and just overexerted himself. Anyway, I'm sure if it was anything more serious, Morgan or Simon would have informed me. I'll check with them when I leave here and let you know."

"Very good."

As Katya left, Sri reminded himself that this was all to be expected. Everyone on his leadership team was inexperienced serving as a leader. The good news was they all had potential. Now, it was up to him to develop that potential. The relatively light duty of a cargo run was an excellent opportunity to return the DSS's faith in him.

Sri turned to Emil. He, at least, had some experience as a line supervisor in Nikko's operations team for the crossing to Eden. Nikko was a fine leader, and he'd expressed confidence in Emil. But he knew this role was a stretch for Emil as well.

"What do you think?"

Emil glanced towards the portal through which Katya left, then met Sri's eyes.

"Obviously, I think you nailed it. But it won't happen by osmosis. And it'll be nearly a year and a half before we reconfig again. The odds of a crisis are extremely low, of course, but should there *be* a crisis, we need her team ready."

"Agreed. We should do some drills. Let's come up with a few scenarios, shall we?"

DAY 5

EMERGENCE

It needed more room.

It was growing. And, as it grew, its proportions changed, which made things worse.

The confinement was becoming intolerable.

It wriggled and turned, yearning to stretch its lengthening limbs.

And now, on top of all that, it was getting *hungry*…

Irritated, bracing its four legs, it pushed against the walls of its chrysalis with its two powerful claws.

It pushed again.

It pushed again, and, this time, it felt something give.

It pushed again, and, this time, it could tell the tip was cracking.

It pushed harder, shifting its footholds, bracing itself, feeling the tip crack even more as it extended one claw and then the other.

Struggling, it turned.

It opened its beak and extended its tongues, exploring the crack.

It pulled and pushed and turned until it could close its beak around a protrusion at the end of the crack.

It pulled and felt the material give.

A piece came off in its mouth, and it tasted wonderful.

It bit more, getting more shards.

Then it began rasping its tongues along the edge, getting more and more of the shell. It all tasted so good, and, as it nibbled, the opening grew and grew.

Finally, it pushed its way out.

It perched, awkwardly at first, on the uneven top of the chrysalis; it stretched its four limbs, one at a time, feeling the joints pop. It unfolded its claws: right, then left, reaching out, snapping them shut. Its eyes were briefly blinded by the light, then habituated as it extended its eyestalks to survey its surroundings.

The physical relief was wonderful!

Then its hunger reasserted itself.

It resumed nibbling at the shell, wearing it down to a nub, as the rest of its senses came alive.

It extended its sensory cluster, its six eyes wavering independently, gathering information, light, color, movement! As its brain processed what it was seeing, two of the eyed fingers monitored the shell as it fed, and the other four moved independently, scanning the area, searching for threats.

Flaps unfurled from the sides of its sensory cluster, stretching to their fullest, channeling sound into its auditory sensors. Chemoreceptive pits collected scents. Nerve endings in the mucus filled sensory pits along its flanks felt air movements as its siblings freed themselves. Its electromagnetic organ sensed the presence of living things. Slowly, its brain processed the flood of information, sorting through it all, struggling to make sense of the cacophony.

Some of it was so *confusing.*

All was new, but somehow, much of it made sense. Much was readily interpretable. But some made no sense; some of the particulars were deeply confusing.

It could sense its siblings. It could feel the warmth of the host's body through its feet. Its chemoreceptors were flooded with the host's scents. It could even feel, through its electromagnetic organ, the pain and anxiety of the host as each chrysalis, having erupted through its skin, tore that skin just a little more as the siblings emerged.

But it could not resolve direction through its electromagnetic organ, nor reconcile what its electromagnetic and gravimetric senses were telling it.

Some things just felt *wrong.*

It felt some anxiety, but more importantly, *much* more importantly, it felt hunger.

Hungry…

Dark colors pulsed along its flanks. Its siblings were emerging. It could feel, through its sensory pits, that some were at least as large as it was. Given the chance, they could be a threat.

It needed to grow. Growth was safety!

Growth required food. Fortunately, there was plenty of food. It just needed to avoid the others and find a safe place to feed.

It walked across the body of the host and, as it distanced itself from the others, its anxiety eased. As its anxiety eased, it could focus on the host's pain, anxiety, and fear.

Good...

It felt the slow, gradual movement of the host's shallow breaths through its feet. It could hear the host's soft moans and whimpers.

So exciting!

The color and texture of the host's flesh changed subtly as it moved across the body. Eventually, as its hunger mounted, it slowed. Safety would come with growth, and growth would come with feasting.

It came across a rounded, red area, with a bump in the center. Slowly, it stretched its claws to grasp the flesh to either side of the bump. As it pinched and pulled, it could feel, through its feet, vibrations as the host drew a quick, sharp breath, followed by a soft, low, guttural sound. It could feel, through its electromagnetic senses, the host's pain. It braced itself and pulled harder. It could tell the pain level was increasing.

Oh, it was delightful!

It opened its beak, extended its tongues to pull the bump into its mouth, and bit down hard. The host vibrated again, then a sharp sound gave way to a soft whimper. As the host's blood pulsed through the wound, it used its tongues to capture every delicious drop and reveled in the waves of suffering emanating from the host.

Waves of bright colors washed over its body.

It drew its tongues across the flesh, rasping tasty tissue. It tightened the grip of its claws and bit chunks of flesh with its beak. As the hole expanded, it shifted its claws to pull more of the flesh, inserted the beak through the hole to tear into the underlying muscle, reveling in the flavors and the vibrations and the sounds and the waves of pain.

Ecstasy!

It was surrounded by what seemed, at least for now, a limitless food supply, a world of fabulous tastes and invigorating sensations as waves of fear and pain came from the host. It could feel itself growing as it fed; its hunger eased but was never fully satisfied. It could sense that its siblings were also feeding, lost in the pleasure of their experiences and, at least for now, not even thinking about feeding on each other.

All was right with the world.

DAY 6

FIRST PREY

Time to leave.

The host was picked nearly clean. The creature's siblings were dispersing. A few of the largest moved a short distance from the remains of the host. They matched the colors and textures around them, then lay in wait, hoping to prey on insufficiently wary smaller siblings. One was feasting now, carefully keeping its victim alive as long as possible. They could all sense its fear and agony as it was slowly, deliberately consumed.

None wanted to suffer the same fate.

The rest carefully moved away from the host. Those that could tear away small pieces did so, continuing to feed as they backed away. Others reached out with all their senses, searching for prey.

The creature had grown substantially since emergence. It was lucky enough, in its initial wanderings, to find itself at the main part of the host's body. It gorged, growing rapidly. Occasionally, a smaller sibling would come too close. Besides offering a little gustatory variety, every sibling consumed was one less threat, one less competitor in the race for growth.

Growth was good. Growth was safety!

It did not, at the moment, sense any nearby siblings large enough to threaten it, but that didn't mean there weren't any. And now that the host was gone, it needed to find other sustenance.

It briefly considered staying to feed on smaller siblings.

At some deep level, some innate programming assessed its chances. How many siblings were there? How many had already chosen this strategy?

The creature decided the better opportunity was to find new prey.

It walked away from the host's remains, then paused, reaching out with all its senses. Everything felt *wrong*, feeding its anxiety. Its gravimetric sensors and electromagnetic sensors weren't providing expected guidance! It needed to get away, it needed to hunt.

But where?

It felt movement of air, then moved in that direction.

As it walked, it could feel a thin material covering a spongy substrate. It matched the color and texture for impeccable camouflage; its six eyes scanned for threats, searched for cover, and sought the best way to the source of the air. It came to an edge and a vertical drop. It walked along that edge, not seeing a path downward that didn't involve a significant drop. Finally, its feet firmly grasped the thin material, and it climbed downward until the material ended. Fortunately, from there, it could sense another horizontal surface.

It withdrew its sensory cluster, pulled in its claws, and let go, splayed its feet. It could feel itself falling. Then, with a soft thump, it hit the horizontal surface and felt a sharp pain from its left midlimb.

It was a hard surface, not spongy like before. The creature stood unsteadily on its hind and midlimbs, claws still tucked. Its left midlimb was throbbing. Something was broken there and would need time to heal, but it could not wait. It reached out with its senses again. Favoring its injured limb, it moved in the direction the air was coming from. Fortunately, with every step, its midlimb healed; by the time it reached the source of the air, the pain was gone.

The air was coming from another vertical surface, this time a regularly spaced mesh that was just a little too tight for the creature to step through. It scanned the mesh, but it didn't see any irregularities in the spacing.

The creature forced its way through the mesh, feeling some give in the material; but, it also suffered more pain as some of its tissue scraped away.

It paused in the strange new space. It could feel the air moving, and the air carried many delectable scents. It moved in the direction of the strongest scent, feeling its sides healing as it walked. The little sensory pits along its flanks detected vibrations in the air. It moved in their direction.

■ ■ ■

The cat froze, its pupils fully dilated. It could hear the soft sound of chewing and knew prey was near. It moved forward slowly, silently, its whiskers guiding it through the narrow spaces of the equipment closet, the faint flow of electronic components generating barely enough light for it to see.

It turned a corner and stared. A cluster of roaches were inside the equipment closet, nibbling on insulation. The cat neither knew about nor cared about the cabling. It had no understanding of the equipment in the closet or the function of the fibers, no realization that, left alone, the roaches were apt to be electrocuted, for the insulation they chewed protected powered connections. The cat didn't understand that human engineers were engaged in an evolutionary war with the roaches. As they continually changed the materials used for insulation, the roaches proved themselves adaptable vermin, continually evolving to find the latest insulation material nutritious.

What the cat *did* know was it found roaches and other vermin intoxicatingly tasty. The product of generations of selective breeding, it was quite adept at catching them.

The cat watched them as they chewed, its body motionless, save the twitching of the end of its tail. It measured the distance, envisioned its leap. It knew that it would only catch one. The rest would scatter. But, it also knew that they tended to return to this place to chew on this insulation. There'd be more chances.

The cat made its choice.

Slowly, it lowered its head, tensed its muscles, then leapt forward, its forepaws landing on one of the roaches, crushing its exoskeleton, causing extensive structural damage. Left alone, the roach would die, albeit slowly.

The cat heard, rather than saw, the other roaches scatter, then lowered its head to feed.

▪ ▪ ▪

The roaches were running!

Each of their bodies ended in two fingerlike cerci covered with tiny hairs. The tiny hairs were connected to nerve fibers, among the fastest of any Earth creature. And the nerve fibers connected directly to their leg muscles as well as their brains. The pouncing of the cat displaced air, which moved the hairs, which sent messages along their nerve fibers. The roaches scattered, running before their brains knew why.

A short distance away, the creature walked cautiously through the strange, alien place. The ground beneath its feet was cold, hard, and, absolutely flat. The air moved steadily, without changing in direction or intensity. It strained its senses, searching for prey, scanning for threats

Suddenly, the pits along its flanks picked up another movement in the air. The vibrations of something moving nearby. It froze and matched the colors and textures around itself.

The roach stopped running.

The roach stood still and began grooming itself. It ran its feet across the surface of its body, then nibbled them clean. It used its feet to pull its antennae to its mouth, running them through its mouthparts to clean them. It did not sense anything else in the ductwork.

Slowly, cautiously, the creature began walking. The light was dim, but even so, its eyes picked up movement. Something was approaching. The creature's auditory sensors picked up the slightest rustling. Its chemoreceptors picked up a new and distinctive odor. Its electromagnetic organ sensed a field generated by something living, something that might be prey.

Slowly, gently, it probed, testing different signals, trying to find something that might confuse the prey.

The roach progressed, then stopped. The roach saw nothing, yet felt *something* was there. It explored with its antennae. They touched the ground, returning information about the texture and temperature, but no scents, no flavors.

Then one of its antennae touched something else. Something it had never encountered before.

The creature remained stock still. The prey was close, so close, yet still out of range.

Long, thin whips from the front of the roach's body were touching everything. One of them touched the creature. It was *so* tempted to strike, to snap out its claws to capture the prey, but it was just a little bit too far away. It kept sensing, to find the right electromagnetic signals.

The creature no longer sensed fear from the prey. The prey seemed to be calming. The creature increased the strength of that signal.

The roach continued to explore with its antennae.

The roach stepped forward to see what its antenna found. Perhaps it would be edible…

The roach took a step, then another.

The creature's claws ached, longing to snap.

Another step.

Another…

Now!

The creature's claws snapped out. One grasped the head just behind the mandibles, and the other snapped shut behind the last pair of legs. The sensors in its claws returned a flood of information about the tastes and textures of the prey. The abdomen was especially enticing. It felt a shell, and it felt folded wings, but it also sensed that the organs were here, as well as the fat stores.

The creature pulled the struggling roach toward itself. It abandoned matching the background, and waves of brilliant colors pulsed along its flanks: reds, purples, yellows. Its vibrant, new patterning reflected the excitement of discovery as it considered how to proceed.

It opened its beak and extended its tongues to explore the prey, reveling in the flavors and textures. Satisfied, it applied the rasping surface of its tongues to hold fast to the prey's body. Then, it bit down with its beak, cutting through the prey's exoskeleton, and exposed the tissues inside it.

The creature felt flashes of pain and fear from the prey. Slowly, carefully, as the prey struggled in vain, the creature stretched its tongues into the body cavity, touching, tasting, exploring, determining which morsels could be torn away and consumed without killing the prey, allowing the prey to live as long as possible, to fully experience every delicious moment of pain and fear while the roach was carefully, delicately consumed.

DAY 7

TIME TO PARTY . . .

Sri sighed as he walked through the corridor.

Everything has plusses and minuses. At 14:37, they'd accelerated enough to preclude communications. It would be nearly three years before they decelerated enough to resume communications.

Twenty-five people would be completely isolated for nearly three years.

Small populations always present psych profiling challenges. They had no room for incompatibilities or annoying habits. And healthy adults had "needs," which meant, especially with such a small population, avoiding people with monogamous inclinations.

Normally, a newly promoted captain returning from a new colony would have little trouble recruiting from the outbound crew, but Eden's allure limited his options. Even with the enhanced bonuses he'd managed to secure, he'd been forced to accept compromises. Not that he was picking up tensions. Well, nothing serious. On the one hand, it had only been a few days. On yet another hand, now that they were underway, they'd be running staggered shifts, which would even further limit the number of individuals interacting at any time.

He and Emil would alternate twelve hour shifts, but most of the crew, including Simon, his physician's assistants, and Katya's three engineering teams would run three overlapping ten hour shifts per day. That would leave Katya and Amar to make their own schedules.

But cargo runs were easy. No migrants, no colonists, no livestock. Not that Sri was fond of animals. He'd enjoyed what he'd seen of Eden wildlife, but such wildlife was always experienced from a distance. He didn't much care for pets and was relieved that his current crew didn't include pet keepers. He didn't care for livestock, either, at least not until served on a plate. But what he *really* detested were vermin. The greasy little bodies of roaches. The buzzing of flies. Stumbling across rodent turds.

He shuddered involuntarily. There was no escaping the filthy little things, of course. They were a fact of life aboard ships, just as they'd been for centuries, dating back to ancient wind driven vessels. They were just too well adapted to their unwilling human hosts, too effective at evading every barrier and exploiting every opportunity. Without livestock, though, there'd be less vermin. The ship's cats ought to be able to keep them under control without having to manage traps. Personally, he'd rather have to manage traps than deal with cats. But, with so few people aboard, best practice was to have a healthy population of cats. Something soothing about them, according to the psychologists. He was just relieved they tended to stay out of sight.

Plusses and minuses…

Sri continued to muse. Without livestock, without agriculture, their food options were limited. He did his best to hoard some real food before leaving Eden, but pickings were slim. Eden's colonial agriculture wasn't sufficiently established to offer much in the way of surplus food. And, of course, the start of a new colony meant no breweries or distilleries. The only alcohol available was moonshine. A particularly good moonshine, but, again, not much in the way of surplus there either. Elke had done her best to help him stock up, but it wouldn't last long, leaving them to synthetics derived from their algal farm until they reached Alexandria. Edible, sure. And nutritious. But not particularly satisfying.

As Sri stepped through an intersegment bulkhead, he realized he could hear music, and then laughter. Isolation parties were a deep space tradition. It would deplete their stockpile of non-algal food and drink, but it helped everyone mark the transition, to acknowledge that, for the duration of the crossing, they were truly alone.

As he got closer, he realized what he was hearing was live music, not a recording. Pretty good music, actually. Deep space crews had long periods of free time to indulge their interests, including music. Sri favored the physical arts, designing and fabricating recreations of the great explorers' wind driven ships. But music was much more conducive for partying!

As he entered, he could see three of Katya's engineers on trumpet, keyboard, and drums. Jim Kuhn, one of Simon's physician's assistants, was playing alto saxophone. Tables were pulled to the side of the room to make room for the musicians and a few people stood around them, moving to the music. The rest were clustered about or grazing the buffet.

He walked through the room, grabbed a plate, studiously selected a few appealing morsels, and poured a cup of moonshine. He responded to greetings with a nod and a smile. Most of the conversations he overheard were retellings of wilderness adventures by those that had been fortunate enough to have them. Holding his cup and plate, he scanned the room until he saw Emil, Katya, and Heinz engaged in a lively conversation, with Amar standing awkwardly nearby.

He walked towards them and, by way of greeting, said, "Amar, my compliments! We achieved isolation right on schedule!"

Amar blinked, then smiled awkwardly and nodded. He was, as always, unkempt almost (but not quite) to the point of insubordination. Not intentional, Sri was sure, but he understood why Elke found him irritating. Still, he was among the best astrogators in the fleet.

Sri raised his eyebrows, waiting for Amar to say something, but Amar nodded again and began tapping at his wrist computer.

"What are you working on, Amar?"

"Oh, uh, I'm, uh, tweaking our course. Before we, uh, reached isolation, I received a transmission with an updated positioning and, uh, it's not a *big* change, but the, uh, revised positioning and associated gravitational field should allow us to, uh, save some, uh, three hours and, uh, forty-seven minutes, give or take…"

He trailed off, blinked, then resumed his tapping. After a moment, Sri turned to see Katya smiling at him.

"He's gone," she chuckled.

Sri turned to look at Amar again. Amar tapped away, his face a mask of concentration, and slowly turned from them.

"Yes, yes he is." Sri shook his head with a broad grin. "Well, it's what makes him so good at his job!"

He turned back, sipped his moonshine, and made a little room to put his cup and plate on the table. He surveyed the party, deciding he should start here, then made the rounds to get better acquainted with his crew.

DAY 8

TIME TO REST...

Sri glanced at the time and winced. It was past 2am and the party was *still* going.

Well, it had petered out to some degree. A number of people trickled out, some after pairing off. He turned to look at the buffet. The real food was pretty much gone, and the moonshine was getting low. He was tempted to refresh his drink, but then he decided he really shouldn't.

The musicians hadn't returned from their latest break. Come to think of it, they'd been on break quite a long time. No matter. Recorded music was playing and that was just fine. He really wanted to leave, but felt he couldn't, not yet. Emil left hours ago. As the executive officer, he'd be on duty by now. Hopefully, he'd managed a little sleep.

Sri sipped his drink.

He'd made the rounds, drumming up small talk: where they were from, what they expected to do with their free time during the crossing, aspirations once they reached Alexandria, etc. At this point, it was all jumbled in his memory. But he'd made the effort, and that was what mattered. He had plenty of time to get to know them all during the crossing.

He panned the room again, then did a quick headcount. A few more were gone. Just a few were left. As he glanced about the room, he caught Yvette's eye. She smiled and raised an eyebrow. He returned her smile, raised his glass slightly, and started walking towards her.

Given the ship's complement, liaison options were limited, especially for senior officers, and most especially for the captain. Some situations were more awkward than others. During the outbound journey, everybody knew about Captain Lubandi and Chief Medical Officer Schmidt; but, while their discretion couldn't hide their relationship, it did minimize the awkwardness. Awkwardness tended to increase with organizational altitude disparity, after all. But, exclusivity (always awkward for such a small group) became particularly dicey with a partner who's not a peer. And captains don't have peers, at least not aboard ship! And, factoring in his preferences further constrained his options.

Still, it was a special night, and Sri really enjoyed Yvette's company. He hoped she hadn't already made plans. He didn't like thinking about her making other plans. It crossed his mind that he ought to do a little soul searching about how much he didn't like thinking about her making other plans. But not tonight!

When he reached her, she continued chatting with Katya despite turning slightly to smile and meet his eyes again.

He waited for a pause in their conversation, then asked, "So, Yvette, I've been meaning to ask you. I didn't see Robert Esposito here. Is he still not feeling well?"

Yvette looked at him quizzically.

"Captain, I don't know what you're talking about."

Eyebrows raised, Sri turned to Katya, who was already waving a hand to get Morgan's attention. Morgan apologized to her companion and walked over, exchanging greetings. Katya redirected the issue to Morgan.

"Morgan, the Captain just asked after Robert. How is he?"

Morgan turned to Yvette. "I've been meaning to ask you. How is Robert?"

Yvette, yawning then shrugging, answered, "Robert?"

Sri raised a finger. "Yvette, Robert Esposito reported that he was too ill to support reconfiguration. He was told to report to Sick Bay. I take it you haven't seen him. Do you know whether he saw Jim or Simon?"

She hesitated, then tapped her wrist computer, then shook her head.

"No, sir. I have no record of Engineer Esposito visiting Sick Bay. None of us have seen him."

Sri turned to Morgan.

"And you, I take it, have not seen fit to check on him?"

Morgan hesitated, turning bright red, then acknowledged that she hadn't after referring him to Sick Bay. She'd just assumed he'd let her know if it was anything serious. Sri paused to collect himself. They were barely underway and already losing track of team members. But this was a teachable moment.

"Leadership is about more than being a taskmaster. We're coaches, mentors. We care for our people. I know you're new to leadership, but I also know you had some fine role models during the crossing from Alexandria."

He glanced from Morgan to Katya, and was confident Katya, at least, got his message. With a nod to Sri, Katya said, "Morgan, I would like you and Yvette to check on Robert."

"Now?"

"First thing in the morning."

She thought briefly about how late it was, then revised her guidance.

"I expect an update from you by 09:00, is that clear?"

DAY 8

GROWTH

The creature continued to walk through the air duct, searching for prey.

It found a few more creatures with exoskeletons. Roaches and crickets were tasty, but they failed to satiate the creature's ever growing hunger. It paused. The sensory pits along its flanks detected fluctuations in air pressure about the same time that its electromagnetic organ picked up a new and different kind of life form. It stepped slowly forward, its sensory cluster held high, its eyes scanning, its auditory sensors wide to pick up the slightest sound.

Finally, its eyes picked up movement. Something was approaching. Something about its size. It froze, reaching out with all of its senses.

The prey's body generated warmth, like the host. Its body was gray and covered with thin, fine growths. The sensory organs appeared to be arranged around its mouth, not in a distinct cluster like its own. That mouth had small, white, sharp pegs, and it would need to take care when striking to immobilize that mouth. There were long, thin projections, apparently sensory, presumably detecting surfaces and air movement, perhaps with chemoreceptors. There were (only) two moist, rounded objects it presumed were optical sensors, and two holes surrounded by flaps that were likely to be auditory sensors. Above the mouth, two moist holes pulsed as it breathed, probably equipped with chemoreceptors.

The rat came closer, and, as it did so, the creature matched background color and texture. It placed its feet, bracing itself for the strike. It began to bombard the prey with electromagnetic signals, using what it learned while feeding on the host. It experimented until it found the right combinations to influence the prey.

The rat hesitated, then stumbled.

Good, good!

It was learning how to read the prey, how to manipulate it! The prey moved closer. It increased its bombardment of the prey, continuing to experiment.

Closer...

Its claws lashed out, one closing around that oral opening, the other grasping its abdomen. The creature held fast, immobilizing the rat's mouth, keeping it closed. Waves of bright colors pulsed along its flanks. The sensors in the other claw reported warmth from the abdomen. It was soft and yielding, and it could feel the beating of its heart, the expansion and contraction of its gas exchange. The organs would be there, as well as a lining of fat stores. Warm moisture filled the floor as the prey voided its bladder.

The panicked, squealing sounds were just delightful. The creature could sense wave after wave of fear and anger coming from the prey as it struggled, but the prey just wasn't strong enough to resist its claws. The creature pulled the prey closer, holding two of its six eyes by the prey's eyes, staring deeply into them, as the other four scanned the prey's body trying to understand its anatomy.

It decided the abdomen was the right place to start.

Staring into the prey's eyes, it used its claw to pull the abdomen closer, then opened its beak. It slowly reached out with its tongues and began pulling and scraping away the little fine hairs that covered the body, feeling the prey's panic and pain.

When it cleared a patch, it slowly, carefully, deliberately bit through the skin. It felt the flesh resist, then tear; it tasted layers of fat and muscle, then seeping blood as its beak sliced through flesh. The struggling and the squealing and the waves of pain and fear were just so, so exciting!

The creature pulled the prey closer, enthralled by the waves of pain and fear, tuning its bombardments, experimenting with what increased and decreased them, then hit the prey with as much sensory augmentation as possible as it took another, deeper bite. It lapped up the fluids oozing from the body. Oh, the fear, the pain, the anxiety wafting from the creature was so delightful!

Still staring into the prey's eyes, the creature reached into its body with its tongues, exploring the prey's anatomy, learning how it worked. The creature considered how to feast without killing the prey, what delectable entrails needed to be left for last so the prey would stay alive. It wanted to share the glorious experience of the prey's consumption with it as long as possible.

And so it fed, and fed, and fed, learning as it went. Until the rat's life processes began to falter. When the prey finally expired, it consumed what remained with efficiency.

And then it moved forward, seeking more prey.

DAY 8

DISCOVERY

Yvette walked briskly towards Robert Esposito's quarters. She was a tad late meeting Robert's supervisor, Morgan. She suspected Morgan might still be wincing a bit after Sri's admonishments.

As she approached, Morgan turned towards her, a bit startled. She did a double take, looking the other way, then back. Yvette blushed, momentarily embarrassed. She realized that her quarters were in the other direction. The two stood just a few feet apart in an awkward silence.

Morgan heard rumors about Yvette's social life and briefly considered asking some pointed questions about whether she'd spent yet another night with the captain, then thought better of it. And, Yvette thought perhaps it would have been wiser to walk the long way around the ring to keep up appearances. Nobody's business, really, so why bother?

Yvette nodded at Morgan, who smiled surreptitiously as she tapped the opaque portal.

"Did you, uh, have a pleasant evening?"

Yvette blushed again, then coyly said, "It was fine, thank you. How was your evening?"

Morgan glanced at her, smiled wryly, shrugged, then tapped the portal again.

No response.

Morgan tapped her wrist computer. "Robert? It's Morgan Filani and Yvette Bauer. Are you awake?"

No response.

Morgan tapped her wrist computer again, then turned to Yvette. "He's in there."

Yvette pursed her lips, then nodded. Sri wouldn't have relieved Emil yet. "Emil, please override Robert Esposito's portal. I need to do a wellness check."

Emil acknowledged, and a moment later, the portal cleared. The room was dimly lit. As they entered, Morgan thought she heard something move.

"Robert?"

No response.

Yvette cleared her throat and said, "lights, full!"

Esposito's living area was a mess. A stack of dishes waited in the sink. Clothes were strewn across the floor. They exchanged a glance. Yvette raised an eyebrow and tilted her head towards the sink.

Morgan shrugged. "I have no idea whether that's normal for him."

Yvette pulled her lips tight, then led the way past the water closet into the sleeping area, then froze.

"My God…"

Morgan, looking back over her shoulder, nearly bumped into Yvette, who was blocking the entry. Carefully, she followed, stepping around Yvette, then stared at the skeleton on the bed.

▪▪▪

"Captain?"

Sri, standing in his water closet and washing his hands, responded absentmindedly.

"Yes? Yvette? How is Engineer Esposito?

Yvette was eerily silent.

He looked upwards, then said, "Yvette?"

After another moment, she responded. "You'd better come see for yourself."

▪▪▪

As Sri approached Esposito's quarters, he was surprised to see Simon coming from the other direction.

"Good morning, Doc. Yvette call you, too?"

Simon nodded, then held out a hand to let Sri enter first. Sri stepped through, then stopped abruptly. Morgan, subdued, didn't acknowledge him. She stood to the side, staring at the space between her hands as she rested them, palms down, on the kitchen counter. Yvette, arms folded, and a bit pale, acknowledged him as he entered. Simon lightly touched Sri's shoulders as he squeezed around him, then paused as well.

Sri wasn't sure how to interpret their body language. After a moment, he managed, "yes? What did you want us to see?"

Without a word, Yvette tilted her head towards the sleeping area. Sri stared at her a moment, then looked at Morgan. He glanced at Simon, then led the way.

Stepping into the bedroom, he saw the remains on the bed and froze in horror. As he regained control of himself, he then stepped through and to the right, beckoning Simon to him. Yvette and Morgan stepped though as well.

Simon, staring at the remains, asked, "You found him like this?"

Yvette nodded, but Simon was still staring at the bed.

"Yvette?"

This time he turned. She nodded again, then said, "exactly like this. We haven't touched anything."

"Are we sure it's Esposito?" Sri asked.

Yvette shrugged and gestured vaguely towards the bed, but could not bring herself to look at the body again. "Who else would it be?"

Simon nodded slowly, then stepped closer to the bed and pulled a small probe from his breast pocket. Using the probe, he leaned closer to look into the mouth. He consulted his wrist computer, then stood and looked over the body.

"Matches Esposito's dental records. The body's the right height. You can see the right ulna has a healed fracture, right where it should be. I'll do a genetic check when I get the remains to sick bay, but I'm confident it's Esposito. Besides, like you said, who else would it be?"

Sri looked at him. "What happened to him?"

Simon shook his head. "I have no idea."

"Disease?"

Simon shook his head again. "I'm not aware of any disease that could cause this. Even if such a disease existed, where would he have caught it?"

Sri turned to him and exclaimed, "Well, *something* happened to him!"

Simon nodded, then began muttering as he leaned forward and examined the remains. "The long bones are all here. Missing a few smaller bones, phalanges…a metacarpal from the right wrist. Bones are all exactly where they should be if his flesh just disappeared."

He stood, head canted to the side as he considered the skull. "The hair! His hair is on the pillow, around the skull, like it just…fell there. Damnedest thing I've ever seen!"

He leaned forward again, using his probe to poke at something on the sheet near the bones.

"That's weird. What the hell…are…"

He stood abruptly. "These are droppings!"

Yvette's voice caught in her throat. "*Droppings?!*"

"Yeah! Droppings. Scats. Poops. Shits. Something, or some *things*, shat on this bed!"

Sri stepped forward and looked squeamishly.

"What? Mice? Rats? What??"

Simon poked the droppings. "No, that's not a rodent dropping. See this white, chalky substance? That's uric acid. Like a bird or reptile would leave."

"Are you saying a bird or reptile was in here? Does anyone on board have any bird or reptile pets?"

Simon shook his head. "No, nobody has any pets. Not as far as I know, anyway."

He looked around the room, then, abruptly, pointed.

"There! There's a carpal, right over there!"

He stepped towards it, and, as he did, something moved.

Simon froze.

"Did you see that?"

Yvette hesitated. Apprehensively, she asked, "see what?!"

Sri stared. He couldn't see anything, but he could *swear* something *moved*. Simon looked about the room and spied a large, plastic pitcher on the night table.

"Hand that pitcher to me!"

Morgan picked up the empty jug and passed it to Simon. Simon, holding the pitcher upside down, slowly crouched by the bone he'd found on the floor. He saw movement again and slammed the pitcher down over it. He leaned closer and suddenly saw a small body with bright colors flashing along its flanks.

Simon heard Sri yell behind him. "What the *fuck* is *that*?!?!"

DAY 8
INFESTATION

The growing creature continued making its way through the air ducts. The roaches weren't satisfying anymore, and barely worth the effort. As it came across them, it stunned them with suppressing bombardments from its electromagnetic organ, then reached out with its tongues to pull them into its beak. Occasionally, it came across small groups, and simply everted its digestive organ through its beak, pressing against them, and absorbed them almost instantaneously.

But they were too small to play with. And oh, how it loved playing with its food.

The rats were still interesting, and, after consuming a few more like the first one, it came across similar creatures that were substantially larger. It stunned them, snapped out its claws to secure them; then, after biting through their bellies, used its tongue to selectively feed. It experimented with its electromagnetic organ, intensifying their sensation, wallowing in their fear and pain as it explored their internal structures. It was learning so much about them! The most recent subject lived a very long time. Just delicious in every way!

But, it was getting harder and harder to find prey. It sensed larger creatures, far too large to be potential prey, at least for now. They were like the host; it wanted them! It wanted so desperately to taste them, to learn about them, to share with them the experience of their torment, such slow, slow deaths. But they were just too big for now. The creature began to wonder whether it needed to leave these dark spaces. It needed to keep growing!

Growth was safety.

The creature could smell the presence of siblings, but not nearby, thankfully. Leaving this place might be risky. It might encounter a larger sibling! But it needed prey, and it had been a while since it sensed any prey. It walked on.

Finally, in the distance, it could see dim lighting. As it approached, it recognized the kind of mesh it passed through to enter this space. Beyond, a much larger space, a darkened space, bore the smells of prey! It was good to grow; but, it realized it would no longer fit through the mesh. It hesitated.

But it was *so* hungry!

The creature extended its eyestalks through the mesh. In the dim light, it could see a wide, open space. It sensed movement, movement that might be prey! Tentatively, it reached out with its claws, grabbing the mesh, squeezing, pulling. It could feel the mesh giving, but not enough. It reached out with beak and tongues, pulling and worrying at the mesh, planting its feet securely, bracing itself.

It could feel the mesh give, and suddenly, it felt the mesh tear! It released the mesh and examined it with its eyestalks. It continued to worry the mesh, pinching and twisting and pulling, until it made a hole large enough to pass through.

DAY 8
ANALYSIS

Sri could feel the tension as he stepped into his conference room.

Deaths happened during crossings, but not like this. There were a lot of ways to die in space, some more gruesome than others, and they'd all seen their share of them. But this…this was different.

He wanted answers as soon as possible; but, of course, Simon needed time to run tests and do some research. Sri allowed him four hours to do so.

Given the gruesome and bewildering nature of the death, Sri demanded the attendance of his extended staff. Normally, he'd hesitate to disturb the sleep cycles of those on staggered shifts, but everybody's sleep cycle was already disrupted by the isolation party. Still, Emil, Jim, and Cleo looked tired.

Can't be helped…

The room was quiet. Everyone stared at a 1 meter cube on the conference room table. Inside, a very strange creature patrolled the edges, probing for an exit. It was about the size of Sri's hand. It had an egg-shaped body with four stout walking legs. Two cruel looking claws tucked under a conical beak. Above and just behind the beak, another odd limb bore six thin, wavering fingers, each ending in a weird, alien eye. Two tails held erect. Bands of reds and oranges flashed along its flanks as it explored the enclosure.

As Sri walked to his chair, the others took to their seats as well.

"Simon? What have we learned?"

Simon rubbed his head with a heavy sigh. "What do you want to hear first? Esposito or the critter?"

"Let's start with Esposito."

Simon nodded and gestured towards Yvette, who tapped her wrist computer. A projection of Esposito's remains appeared next to the cube, eliciting groans and gasps.

Yvette grimaced. "I'm sorry. I know it's not pleasant." She stood and walked towards the projection.

"We confirmed this is in fact Esposito. Genetics matched." Pointing at the projection, she said, "Next question is what happened, and…I haven't a clue. I've never seen anything like it. Can't find anything in the literature, either. It's like he just laid back and, what, dissolved?"

She shook her head. "Not dissolved. That would imply some residue, and we found none. Not much blood, either. Evaporated, maybe? Yeah, that'd be a better word…"

"How did he die?" Sri asked.

Yvette stared at the display, torn between professionalism and panic. She'd seen death before, hideous death. But nothing, ever, like this. She stared, aware of the awkwardness of her silence, and, at least for the moment, not caring. She hadn't signed up for *this*!

It was too much. Just…too much.

Evaporated tissue?

DROPPINGS?!?!

She shook her head, collected herself, and turned to look at Sri.

"How did he die? I have NO idea how he died! The soft tissues are all gone, right? Without the soft tissues, it's nearly impossible to say. Injury? Wound? Organ failure? Toxins? If a toxin was involved, it wasn't in his system long enough to travel to his hair or bone tissue.

And there's nothing in the positioning of the bones to suggest anything. The sheets are smooth. Apparently, he just calmly laid down and…died? We checked monitoring systems. If there'd been any quick, sudden movements, anything consistent with violence or any kind of struggle, they'd have maintained a recording. *Nothing…*"

She returned to the projection and sighed deeply. "The bones are clean, like they were dipped in some sort of solvent, then carefully put back, as if that made any sense."

She paused, turned, then, with a grimace and a slightly softer voice, added, "And some of the bones look scraped or rasped."

Morgan, her face betraying her revulsion, interrupted.

"Are you saying something *ate* him?!"

Yvette turned abruptly toward her, as if she felt accused.

"Fuck, Morgan, I don't know! That would explain the disappearance of the soft tissues, right? Makes more sense than some mysterious villain dipping his bones in solvent! Then we have this little bastard, right?" She gestured rapidly towards the cube. "So, yeah, maybe something ate him!"

Sri rapped his knuckle on the tabletop to regain control. The inexperience, nay, immaturity of his crew was taking control, rapidly. He had to admit some empathy, but he couldn't allow it, and it was now his responsibility to lead.

Simon sighed, then stood. "Okay, it's time to talk about our guest."

Everyone focused on the creature pacing inside the cube. With a glance around the room, Yvette took her seat as Simon began pacing around the table.

"Let's start with the obvious," he began. "For anyone that hasn't figured this part out yet, that thing ain't from Earth. I'm no xenobiologist, but I remember enough comparative zoology to tell you that much! So, if it's not from Earth, then it must be from Eden, right? Because, well, where the hell else would it be from? And at those 'show and tell' presentations before we left, the xenobiologists showed us some critters that looked a lot like our little friend here. Which raises some other questions."

He stopped and held up a finger. "First, if it's from Eden, how the hell did it get into Esposito's quarters? Let's think about that for a minute! Eden's xenobiology labs are in their landing craft colonial hanger decks. Makes it easy to transfer specimens to the lab, and, if there's a breach, you just open the doors to let the vacuum of space sterilize the hanger deck…"

He waved his hand and shook his head, then continued. "None of that matters right now. Here's what matters. If that thing came from Eden, then let's think about just how improbable a journey that would be!

It didn't just escape a xenobiology lab. It, somehow, managed to get from the Eden landing craft hanger deck to the hanger deck for our shuttle tug. It hitched a ride on our shuttle tug, without any of us noticing, mind you, to the *Lucky Strike*'s hanger deck. And, then! Then it managed to make its way from the shuttle tug hanger deck at one end of the ship's drum through thirty-eight rings of cargo pods to the residential ring and Esposito's quarters!"

Simon glanced around the room, eyebrows raised, then continued.

"It gets worse!"

Simon tapped his wrist computer. The projection zoomed out to show Esposito's sleeping area. He tapped again, and small shapes were highlighted.

"We found droppings in Esposito's quarters. *Lots* of droppings on his bed and on the floor. So, either our little friend poops a LOT, or we have more than one guest. We've observed this one for a bit over four hours. It's only pooped once. That's not a particularly solid data set from which to extrapolate, but, if we presume the things poop once every four hours, we figure whatever happened to Esposito took place immediately after the last time he communicated with anyone. Assuming this, it looks like we have, give or take, about twenty unwanted guests."

He held up his hands and shook his head. "So, twenty or so of these little critters somehow snuck aboard our shuttle tug, crossed over with us, and then, for some reason that I can't fathom, made their way, sticking together, through the ship to Esposito's quarters. Nobody else's quarters; just his. Why? And where are they now?"

Katya shook her head. "No, that doesn't make any sense! Remember, when we made the crossing, the drum was rotating. Then we launched and reconfigured. If these things were making their way from the shuttle tug bays to Esposito's quarters, wouldn't we have noticed them during the reconfig?"

Simon met her eyes, shrugged, and looked around the room. Then he resumed pacing as he held up two fingers.

"Second, you may also remember the xenobiologists talking about biochemical incompatibility. They were pretty excited about that, remember? Because it meant we might be able to live on the surface of Eden without domes or environmental suits. Because biochemical incompatibility would mean we wouldn't have to worry about Eden germs or weird alien diseases! It also means we can't eat anything from Eden. No fruits, no veggies, no meats. And nothing from Eden should be able to digest anything from Earth..."

He stopped pacing and looked around the room, making eye contact with each of them as he did so.

"I analyzed a number of droppings from Bob's quarters and one from our little friend here. They *all* contain DNA from Bob!"

He paused to let that sink in, then resumed pacing, his steps and figure becoming more erratic and fearful. "Third, and finally, *how*? How were these things able to feed on Bob? Was he already dead? If so, how? If not, how did they subdue him? Venom? If we were talking about some mysterious Earth creature, you'd think venom, right?! But we're not. We're talking about an alien life form, and venoms are complex chemicals. They should not be able to digest Bob, and they should not be able to produce a venom that would work on humans."

He stopped, shaking his head in disbelief. "Which is impossible! They're aliens, right? They should NOT be able to digest anything alien to their own biochemistry! It's just not possible!!"

A heavy silence hung over the shaken crew. The only sound in the stillness was the quick, hyperventilating breaths of several officers. After a moment, Sri asked, "so, where does that leave us?"

Simon barked a laugh, then exclaimed, "fuck if I know!"

Emil started to stand, but Simon, shaking his head, waved him back.

"I apologize, Captain. I'm just…at a loss!" The room was silent as Simon returned to his seat.

"If there's more of them, where are they?" They turned to Amar, who blinked and repeated his question.

"Very good question!" Simon responded. "As I said, there were a *lot* of droppings in Esposito's quarters, mostly on his bed, but also on the floor between the bed and the air vent. That suggests the things went into the ventilation system."

Heinz shook his head. "There's no way that thing could fit through the mesh covering the air ducts!"

Cleo, lips pursed, countered. "Unless they were smaller when they did so. Which would suggest they didn't all go at once. But it also begs the question of why this one stayed."

Simon nodded. "Well, here's something to consider. As I said, I analyzed one of its droppings and found some of Esposito's DNA. I also found alien DNA. Nucleic acids, but not the nucleic acids shared by all Earth life. Presumably Eden DNA. Where'd *that* come from? Well, either our guests packed little Eden lunches, or this one turned cannibal. Which might explain why the others went into the vents, to flee from this one."

The entire room shuddered at the thought.

Sri nodded, then stood. "Okay, we have some mysteries, but we need to table them for now. When we reach Alexandria, we'll consult with their xenobiologists, and talk through the whole biochemical incompatibility thing."

He looked around the room. "And I don't know, or for the moment care, how they managed to reach Esposito's quarters. All I care about is whether anyone else is at risk. If they're in the duct work, then they can move freely through the command and residential rings. If they can subdue Esposito, then we're all vulnerable."

He paused again, then continued. "I don't want anyone alone until we figure out how to find and kill these things. Let's put cots in the breakrooms so people aren't sleeping alone. Let's make sure people on duty aren't alone, either. Emil, begin assigning shift partners."

Nodding around the table. He turned to Katya.

"We can't fit in through the ductwork, and I don't think we want to wait for them to come out. We need a way to find them and kill them. Get your team together, come up with some options. Let's reconvene in four hours?"

DAY 8

TEAMWORK

The creature moved through the broad, open darkness.

The wounds it suffered squeezing through the ragged opening in the mesh ached as they healed. This place felt even more wrong than the other side of the mesh. The substrate remained uniformly flat and cold to the touch, but there were no vertical surfaces, at least not yet. Structures interrupting the void, yes, but no vertical surfaces, no boundaries. Somehow, especially in this darkness, that made it feel even more exposed. The colors flashing along its flanks, though not visible in the darkness, reflected its anxiety.

The creature did not, of course, understand where it was nor where it had been. The ductwork supporting the residential spaces extended through this broad, open, interstitial space to the life support core. When the ship was in port and the drum rotated to simulate gravity, these passageways extended upwards, through the ceilings, to the core. The sense of gravity eased as one climbed, with the core itself experiencing weightlessness. But, now, as the ship relentlessly accelerated towards the midpoint of their journey, the drum was not rolling. "Down" was aft, and this interstitial space was a broad, empty, dark wilderness frequented by only the ship's vermin and the cats hunting them.

As it moved forward, it reached out with its senses. There were rodents and insects to prey upon, but not many. It did not, could not understand that, since humans rarely entered this space, there was little human detritus to sustain vermin. Those present were exploring in transit, evading a cat, but could not stay very long without sustenance. It could also sense the presence of something else: larger, massive creatures like the host. It had sensed these things while in the ductwork but managed to avoid them. The prospect of encountering them here, without shelter or vertical surfaces, fed its anxiety, but none seemed close enough to worry about. It continued to explore.

Suddenly, it sensed the presence of another of its own kind.

It paused, concerned. It couldn't tell much from here. Was the other big enough to be a threat? Small enough to be prey? A sibling? The safest course of action would be to move away, but it was hungry. So, so hungry!

Cautiously, it moved forward, reaching out with all its senses, desperately seeking information about the other. The prospect of breeding didn't even occur to it. It was not yet time; they were both far too small.

As it moved forward, it could feel, through its electromagnetic senses, the other was also hungry and cautious. It realized they were about the same size. What a quandary!

If they engaged, the winner would secure an ample meal, but it would likely be too damaged to survive another challenger. If prey were abundant, they'd simply avoid each other. But, prey was so scarce, hunger and the presence of a similarly sized and hungry sibling activated new programming.

It was an intelligent creature, but not a reasoning creature. It had neither abstract thought nor language. It was the ultimate pioneer, exquisitely adapted to exploit ecological disruptions. Its spores infected creatures with stressed immune systems generated by ecological disaster, be it fire, earthquake, or tsunami. Crafted by evolution to emerge and feed and grow as rapidly as possible, they generated and spread as many spores as possible in anticipation of the next disruption on a highly dynamic world.

Its behaviors were a broad portfolio of strategies for nearly any conceivable scenario. It was a complex web of preprogrammed, instinctual responses to a broad range of potential environmental circumstances, with the capacity to fine tune through its intelligence, through its ability to experiment and learn. Hungry, sensing the other, weighing the scarcity of obtainable prey, its programming shifted.

It moved towards its sibling, and, as it did so, began to emit a new pheromone. As it cautiously moved forward, it detected compatible pheromones from the other. That was good.

It changed the textures and colors pulsing along its flanks, then saw the other was also displaying the new colors. Cautiously, carefully, claws slowly spreading, they approached each other. Staring into each other's eyes, the tips of their claws touched; they each moved left, circling each other. They opened their beaks, stretched out their tongues, tasting each other, verifying their common genetics. They recognized each other as siblings. Finally, they stopped, withdrew their claws, closed their beaks, and shifted their displays to match the colors and textures around them.

Together, they began to hunt.

DAY 9

TACTICS

Sri returned to his conference room a little early and found Morgan, Heinz, and Katya engaged in a lively discussion while pointing at a wireframe schematic hovering over the table. Cleo sleepily looked on as she absentmindedly scratched behind the ears of one of the ship's cats; it purred softly and fed vigorously from a small bowl beside her on the table. The cube holding their mystery guest was on the floor, pushed to a corner. The rest of the team trickled in behind him.

Something about the cat bothered Sri, but he couldn't quite put a finger on it. He shrugged, turned to focus on the wireframe. "That looks promising!" he said, not to anyone in particular.

Morgan and Heinz stretched their backs as Katya said, "We think it'll work."

"Walk me through it, please."

Katya rubbed her neck from stress and exhaustion. "Okay, we think they're in the ductwork, right? Because of what Simon said about the droppings, and because nobody's reported any encounters with anything unusual. Unfortunately, we're too big to chase them through the ductwork."

Sri nodded. Then, as his mind clicked, Sri turned to stare at the cat.

"Cleo, what's that cat eating?"

She blinked, then answered. "Synthetic chicken, why?"

"Because it tastes like shit, that's why!"

They'd run out of real food at the party. All they had left for the rest of the journey was synthetics fabricated from the life support system's algal tanks. Nutritious? Sure, but not particularly appealing. One of the many reasons inbound crews tended to spend a year or so exploring a colony before signing on to another crossing was to indulge in real food, get the taste of synthetics out of their mouths.

Simon snapped his fingers. "You're right, it tastes like shit. And the cats are bred to prefer vermin anyway. If that cat's eating syntho-chicken, it must be starving!"

His voice trailed off as another cat wandered into the room. It mewed forlornly, then began purring as it rubbed against Heinz's leg. As Cleo sighed and stood to fetch another bowl, Simon suddenly realized the implication.

"The critters must be hunting the vermin!"

Katya interjected. "So, after eating Robert, they're suddenly satisfied with vermin? They didn't notice any vermin as they made their way to Esposito's quarters?"

"Not going through the interstitial spaces," Heinz answered. "Remember, there's no ductwork in the cargo rings. They had to go through interstitial spaces between the hangar bay and the command and residential rings!"

Katya, leaning back and staring into space, pondered that for a moment. Standardizing drum diameter at 500 meters was a compromise. In port, the drum spun to simulate gravity, so "down" was the outer edge of the ring and "up" was towards the center of the ring. Moving "up" meant moving towards the center of the ring, with one's sense of gravity declining along the way until reaching zero-g at the center. Living and working space for the command and residential rings were along the outer 20 meters of those rings; the interstitial space, the space between the livable outer 20 meters and the center of each ring was used for engineering infrastructure and storage.

A smaller diameter would reduce the usable space along the inner edge of the ring, requiring more rings for the same net area. A smaller diameter would also require rotating the drum at a higher rate to simulate 1g. But crossings involved relentless acceleration, then deceleration, so "down" was aft, then forward. During a crossing, the interstitial space offered the same simulated gravity experience as the outer edge. A wider ring meant more interstitial space, and they had more than enough already. Even at 500 meters, that came to over 160,000 square meters per ring and, including the cargo rings between the shuttle hanger decks and the command and residential rings, there were twenty rings!

She turned back to Heinz. "Okay, I'll buy that. It's possible they could make their way through interstitial spaces without encountering any vermin. But we should remember that we *do* have vermin in there, and the cats hunt vermin in the interstitial…"

Sri abruptly slapped the table and barked. All of this talk about vermin was nauseating. "Enough about the vermin! It doesn't matter! They're in the ductwork, and they're eating vermin. How do we find and destroy them?"

Katya, ego bruised, struggled to collect herself. *Asshole! I shouldn't be here… I should've stayed on Eden…*

She turned to the projection, pointing as she spoke. "Uh…okay, so, we're designing this gizmo to hunt the critters. This is actual size, so it'll fit through even the tightest spaces in the ductwork. The four wheels are big enough to let it roll over any irregularities, and these robotic arms are long enough to reach anywhere within the ductwork, even the wider and taller junctions. The arms end in a grasping pincer to hold onto our little guests and a little torch to incinerate them."

"Kinda reminds me of our guests! How does it find them?" Sri observed.

Katya pointed to Morgan, who said, "Well, that's the challenge, right? Go look in the cube."

Sri stood and walked over to the enclosure, which appeared to be empty.

"Where is it?"

"Exactly! Trust me, it's still in there, but its camouflage is amazing."

As she spoke, she walked over to the enclosure and gave it a firm shake, then another, and another. On the third try, one of the sticks crisscrossing the interior of the cube fell to the floor. As it did so, the creature suddenly appeared, bright colors flashing along its flanks. It righted itself, backed up a few centimeters, then disappeared.

Sri bent over to look more closely. As he stared at the spot and moved slightly side to side, he began to make out the shape of the thing. But only barely! It matched the colors and textures of the cube's substrate perfectly. He turned to Morgan and exclaimed, "amazing!" Then he turned back. But, this time, he couldn't find it.

Morgan nodded. "Yep, and a real problem. We started the design presuming a camera and motion sensors, but clearly we can't count on the camera. We'll keep it, so we can see what the gizmo sees, but it won't find our little guests visually. The motion sensors will work, but from what we've observed of our little friend here, once they settle down, they don't seem to move much."

She hesitated, then added, "We're struggling here a bit, right? Our new pet doesn't move much, so we're assuming our unwelcome guests aren't moving much, either. Why? Well, if you're right and they're munching vermin, perhaps they operate as ambush hunters?"

She shook her head. "I'm worried. Sure, the gizmos will work. But, we're making a lot of assumptions here, and I'm worried that we're missing something. Why did these things attack Robert, and how? Why haven't they attacked anyone else? I don't understand why they would attack him and only him! Why haven't we seen any of them? Are they in the ductwork? Or in the interstitial space? Or did they go dormant or transform or, hell, I dunno, somehow flip into some alternative universe."

Katya walked over and placed a hand on Morgan's shoulder. As Morgan sat, Katya turned to Sri and said, "It's been a rough morning."

Sri, suddenly embarrassed by his earlier outburst, nodded and said softly, "I understand."

"Okay," Katya continued. "While I share your discomfort, Morgan, I don't know what to do about it. All we *can* do is strive to be open minded, open to the possibility that we're missing something. Beyond that, I can only deal with what's in front of me, right? And what's in front of me is the need to catch and kill these little bastards."

She tapped her wrist computer and the room went dark. "We can't count on the camera or the motion sensors to find them. Next thing we thought of was infrared."

A small orange glow appeared in the center of the cube, presumably climbing another stick. She tapped again, and the lights came back on. "Similar to an Earth mammal or reptile, they have a heat signature. That helps! We'll use a camera that can pick up infrared." She turned and pointed towards Heinz. "And, Heinz had the bright idea to test the air in the enclosure. Turns out they have a unique chemical signature, so we're designing our gizmo with a robotic nose. So, the gizmos will roam the ductwork to pick up and follow their scent. Then, when they're close enough to pick up their heat signatures, they'll grab the buggers with this pincer and incinerate them with that torch. Questions?"

The room was quiet, so Sri asked, "When can they be ready?"

"Well, that's where we have another challenge. On a run like this, we don't normally need to fabricate much. If we do, it's usually something solid to patch or plug a leak or, if we need to replace a failed part, well, they're all standard parts, right? For this, we'll use standard components where we can, but there will be custom parts that need to be designed as well as fabricated. Bottom line? Morgan and her team should have a prototype ready within about five hours. We'll see how well it works so we can refine the next one as we build it. Then we can probably churn one out every two hours or so."

"Very good. Get started. Keep me posted!"

DAY 9

AMBITION

The creatures roamed the broad, dark space.

The largest took the lead; the others explored behind it. The pack's movements seemed random, pausing here, poking there, but always within a few body lengths of the alpha. Their eyes were nearly useless in the dark. They held their eyestalks and sensory clusters near their sheaths for protection, but both were ready to extend for those rare patches of dim lighting encountered here and there.

Their other senses periodically alerted them to the presence of the little crunchy things and hairy things that had sustained them thus far. The creatures heard them, or smelled them, or felt their movement through the sensory pits along their flanks. Individually, as they encountered them, the creatures suppressed them electromagnetically. Thoroughly stupefied, the now disappointingly tiny prey remained immobile, radiating fear and anxiety, as the creatures extended a tongue to snatch them up and pull them into their beaks. They remained nutritious, and their fear and anxiety was pleasing; but, as the creatures grew, such prey were too small for experimentation. Growth was their imperative; growth was safety. And, increasingly, they sensed that something even more desirable would be attained with adequate growth.

Meanwhile, available prey were just so disappointing. Boring and small. At some point, such prey would simply not be worth the effort to consume. They needed something *bigger*.

Occasionally, as they moved together, they encountered more of their own kind. Smaller individuals were quickly dispatched and feasted upon. Larger individuals, once their common genetics were verified, were invited to join the pack and linked, constantly, through their electromagnetic senses. As they grew, their linkage intensified at the expense of their sense of self. At times, they could no longer distinguish their own thoughts from their siblings. Their aspirations gradually, subtly shifted from growth as individuals to growth of the largest, especially the alpha. They *must* ensure at least some of them grew large enough, though they did not, as yet, understand what they needed to be large enough to achieve. But increasingly, when prey encounters lagged, when the hunger pangs of the largest intensified, when growth was at risk, the smallest did not object when their larger siblings fell upon them, dismembering and consuming them in order to sustain growth.

There were seven of them now. Seven roaming creatures, reaching out with their senses.

The leftmost paused, then informed the others that it felt a new presence. They froze, reached out, unable to discern the new presence themselves, but able to experience it through their alpha sibling. The leftmost moved cautiously towards the presence, picking up more information as it did, and shared that information with its siblings.

The presence was moving towards them. It felt similar to the little hairy things, similar yet different. It moved stealthily, as if it might be hunting them.

How exciting!

They spread out, forming a broad, shallow arc. Despite the darkness, they matched colors and textures with their surroundings, as best they could determine them, and moved slowly towards the presence. They continued to explore it with their electromagnetic senses, gently probing, experimenting, applying what they remembered from the host and what they learned from the little hairy things.

Their other senses began to report information.

It smelled a bit like the little hairy things, but it also smelled like the crunchy things. They began to realize it was also a predator, that it also fed on the little crunchy and hairy things, that what they were smelling were the residual odors of *its* prey. This was exciting, but merited caution. A predator would almost certainly fight back. They'd need to thoroughly suppress this creature; not only subdue it and feast upon it, but they needed to experiment, learn about it, and be ready for future encounters with its kind. But they never even considered passing by. It was large, larger than any one of them, and that meant it had the biomass they desperately needed to fuel their growth.

They continued to bombard it, realizing they were approaching a dimly lit space. They decided they would take the prey there, taking advantage of the opportunity to visually inspect it as they consumed it. The alpha and the creatures beside it stopped moving forward. The others extended the arc to flank the prey as it approached.

There! It was large! It was covered with hairs, nasty, nasty hairs that would need to be removed. Its optical sensors were forward, to help it detect its prey. Its air intake was also forward, no doubt filled with chemoreceptors. It had sensory whiskers about its snout, and a mouth filled with sharp teeth. The legs did not appear to end in claws, which would be surprising. Or did they? Might there be sheathed claws, claws that could be extended?

The creatures could sense which signals effectively suppressed it, learning as they experimented. It continued forward, stumbling occasionally. It closed its eyes and shook its head; then, foolishly, the new prey continued forward.

As it approached them, the second largest sibling scurried forward, snapping its claws forward to hold the mouth shut. The others struck, as selected pack members immobilized its legs, and the rest struck the prey's midsection. Their combined weight was too much for the prey. Coordinating their efforts, they pulled the prey down. They opened their beaks to extend their tongues, tearing away the nasty little hairs to expose the prey's skin. When enough flesh was exposed, the alpha, having held back to ensure the safety of their largest pack member, moved forward.

They held the prey down and gradually eased suppression, balancing their desire to enjoy the prey's anxiety and coming agony against any risk of its escape. The alpha slowly, deliberately reached forward and grasped the now naked flesh of the prey's belly. It opened its beak and slowly bit deeply into the flesh, feeling the fat and muscle under the skin, reveling in the flavors of the meat and juices. It sensed the panic of the prey as well as its pain.

The alpha extended its tongues through the opening, exploring, tasting, selectively taking morsels of meat and fat that, based on their experience with the smaller hairy things, could be taken without killing the prey. It found the prey's nerve cord and paused its feeding. It explored the nerve cord to sense which fibers carried pain signals back to the brain and which carried motor control signals to the voluntary muscles. Taking great care to preserve the pain signals, the alpha worked its way forward, towards the prey's brain, and teased out the motor controls. With a quick snip of its beak, it severed them.

The prey's struggles ceased.

The mouth and legs went slack, yet they could all feel and enjoy waves of pain and anguish coming from the prey. They released their grip and began carefully dissecting their prey, tearing away the cat's fur, exposing its flesh, and feasting selectively to keep the prey alive as long as possible. The pack continued to experiment with electromagnetic suppression, key learnings for hunting such prey in the future. They also experimented with sensory augmentation, maximizing the prey's pain and fear as they joyfully shared the experience of the prey's consumption with it. When the prey finally expired, the creatures consumed what remained, growing as they fed, knowing they wanted to find more of these delicious, delightful prey.

DAY 10

FIRST DEPLOYMENT...

Katya and her team surrounded a gray bubble atop the conference table. Within the bubble they could make out a highlighted duct crawler. Despite not moving relative to the table, they could tell it was rolling forward through the ductwork, as landmarks such as scuffs and debris moved along the tabletop. Above the projection were a set of gauges, and one needle began moving.

Katya pointed at the gauge and exclaimed, "look! It's picked up a scent!"

The gizmo periodically paused. Two sensory probes wavered left and right. Then, it started moving forward again, the needle slowly, steadily increasing its speed. It reached a juncture, then stopped to move the sensors right and left again. As it did so, the gauge needle moved. After a moment, the gizmo turned left and continued. Suddenly, it stopped. Half a meter in front of it, a shape glowed, a silhouette very much like their guest.

Katya, without thinking, whispered, "It sees one! The thing isn't moving; it's probably confident in its camouflage. Only reason we can see it is because we're using infrared."

One of the gizmo's robotic arms slowly, silently extended until it was over the glowing shape. Suddenly, the arm's claw dropped and snapped about the creature, then lifted it from the substrate. The gizmo turned on its lights, and they could see the creature as its limbs flailed, claws snapping futilely, and flanks pulsed red and black. It was about the same size as the creature in the enclosure in the corner of the room. The gizmo's other robotic arm extended a glowing tip towards the captured creature. When it was within range, the glowing tip ignited, incinerating the struggling creature with a short burst of flame. The room erupted in applause as ashes fell to the floor, out of camera range. The gizmo retracted its arms and turned off its lights. As it moved forward into the darkness, the gauges above the image registered the faint scent of another creature.

DAY 11
SCARCITY

The pack hungrily explored the dark space.

They opportunistically snatched up roaches, rats, and cats when they could, but they were getting rarer and rarer. And, as the creatures grew, the ship's vermin just weren't worth the effort. Their sense of urgency was increasing. Growth was safety. And they were, indeed, growing. The alpha was bigger than the cats they hunted.

But, they'd started picking up alarm pheromones. Somewhere, not too far away, something was killing their kind, killing them violently. At some instinctive layer, this information was factored in.

The pack's priorities changed.

Growth remained a priority, but growth was now a means to an end. With growth would come maturation, and with maturation would come mating, and with mating would come offspring. As siblings, they *must* ensure that at least the largest grew enough to mate, to produce offspring. Preparing to produce offspring became their top priority.

They weren't big enough, not yet. But as they smelled the awful smell of death, somewhere deep inside their bodies, changes occurred. Changes they did not yet understand; changes that would reduce the growth threshold for their sexual maturity.

As long as they continued to find ample prey, the seven would work together. But should their food supply dwindle, in order to ensure the successful passing of their genes to the next generation, the smallest would willingly sacrifice their lives.

DAY 11

THE HUNT CONTINUES

Sri yawned. There were now three gray bubbles atop the conference room table. The original gizmo, idling in the center, had destroyed six of their unwanted guests, according to the counter above it. To its right, the second gizmo, deployed almost four hours later, accounted for three more. To its left, the third gizmo, deployed almost two hours ago, hadn't sniffed out any yet. Neither of the first two reported detecting any sign in nearly two hours. He looked about the room. Everyone was exhausted, especially the ones that should be asleep right now.

Time to reconsider strategy.

"Simon, your original estimate was about twenty of the things. Between the nine killed by the gizmos and your pet, that's ten. What do you think? What's your confidence in that original estimate?"

Simon stretched and stifled a yawn, then scoffed. "My confidence? Low. Very, very low. Do I think we've killed them all? No. Do I think there's another ten? No idea."

Sri turned to Katya. "What do you think?"

"I'm with Simon. I don't think we've nailed them all yet, but we really have no idea how many more there are. I'll say this, though. I don't think there are any more in the duct work."

"No? Where are they, then?"

"Can't say for sure," stifling a yawn, Katya shrugged. "But I'd guess the interstitial space. I mean, really, if they're not in the duct work, and nobody's seeing them, then where else would they be?"

Sri nodded again, then he leaned back and stared at the ceiling. After a few minutes, he leaned forward and splayed his hands. "Time to revisit strategy. Let's start with this. How can we determine whether we have any more guests?"

Heinz raised a finger. "Well, I can increase the sensitivity of one of our sensors. It won't be able to tell us how many or in which direction, but, if there are any left, it'll smell them."

Sri hesitated, then asked, "What about Simon's pet? How do you isolate it?"

Simon waved a hand. "No worries; it died. Starved, I'd guess. Could not find anything to feed it. And it wouldn't take any of the synthetics we tried. Anyway, it's on ice now."

"Ah…too bad. Was useful having one for testing. Heinz, proceed. Next?"

"I don't think there's much point in building more gizmos…" Katya observed.

"Agreed. Can they keep hunting? Set them on auto? Have them notify us if they find any more?"

"No reason not to, right?"

Cleo raised a finger and asked, "Are we being too cautious?"

"What do you mean?"

"The gizmos haven't found any much bigger than Simon's pet. Nobody's seen any of them anywhere else. Nobody's been bothered by them. Are we being overly cautious?"

Morgan scowled. "They killed Bob! And they were even smaller then!"

Cleo raised a hand. "I know, I know. We have no idea how they managed to do that. Somebody said Bob was sick, right? Maybe that's what it was. Maybe he was really, really sick, maybe he couldn't fight back or something. Hell, maybe he died, and they fed on his corpse. Dammit, I don't know! But we've killed nine of the things and, so far, they're pretty unimpressive. Pretty easy to kill." She looked around the room, then added, "How about we make some handheld incinerators and send our engineering teams hunting?"

Sri leaned back in his chair and looked at Emil, who shrugged. "I can't argue with that," Emil said thoughtfully. "They don't seem threatening; nobody else has been bothered. I don't know what happened with Bob, but maybe it's not as bad as we initially feared. Maybe you're right, he was just really sick or already dead when they found him."

Simon shrugged. "Can't say, right? Wasn't enough of him left to know. But that would explain why there was no sign of struggle."

Sri nodded, then asked, "You think they're in the interstitial space? Why not deploy the gizmos there?

"Well, we could but they're designed for the ducts," Katya responded. "The interstitial space has a lot of broad, open areas. Almost 170,000 square meters! I suppose we can reprogram the crawlers to set up a search pattern." She paused, then added, "There's also structural framing, and Simon's pet climbed pretty well. Gizmos aren't designed to deal with that."

Sri sighed. "Okay, leave one of the gizmos hunting in the duct work. Reprogram and redeploy the other two to search the interstitial spaces. Don't know how successful they'll be, but they're already built so we might as well use them. Start fabricating handheld incinerators and we'll start hunting them as soon as they're ready."

Katya glanced upward, tilted her head side to side a few times, and said, "Gimme a couple hours to reprogram the gizmos. We'll start fabricating the incinerators right away. Should take us about, oh, an hour or so per unit."

Sri stood. "Make it happen. And let's get everyone back on their normal sleep cycles. We're no good if we're zombies." He glanced at the time, then continued, "Heinz and his team should be sleeping now. Morgan, take over the fabrication center. Switch to incinerators so Cleo's team can hunt the little buggers. End of shift, Heinz, you'll take over fabrication, Morgan's team will hunt, and it'll be Cleo's turn to get some rest for her team. And, Katya, make sure you get some rest yourself!"

DAY 12

ENCOUNTERS ...

Donna stood in a breakroom near the captain's conference room along with the rest of Heinz' team. They were all exhausted. They'd caught a bit of sleep here and there. Cat naps at best, not very satisfying given the background noise. It was especially frustrating because there really wasn't any reason to be awake. In theory, they were supposed to be running fabricators with Morgan's team, but there really wasn't enough room. This left her and the rest of Heinz' team idling in the breakroom.

Donna understood the motivation of keeping everyone awake. Safety in numbers, but nothing was happening. So, when the word came down that they were to return to their staggered shifts, Donna let out a very, very weary "whoop."

She hesitated. They were supposed to use the uncomfortable breakroom cots so nobody was alone, but nobody was using them. Screw that! She wanted to sleep in her own bed!

Donna stumbled towards her quarters, managing a reasonably adept walk through the corridor, but, once through her portal, all bets were off. She kicked off her shoes, loosened her clothes, and may have lost consciousness before actually landing on her bed.

■ ■ ■

The pack was hungry.

They hadn't come across a cat in quite a while. They hadn't seen any roaches or rats, either, but that didn't matter. They were just too small to bother with. They didn't detect alarm pheromones anymore; but, then again, they couldn't detect any sign of their siblings. It seemed increasingly likely they were the only remaining survivors, and whatever killed their siblings would start hunting them now.

Deep inside their bodies, more non-verbal decisions were made with respect to the strangeness of this place and the apparent hazard to their kind. Growth and breeding were imperatives, if not for all of them, then for at least one of them. One of them *had* to breed! But growth required prey, and there was no prey.

Then, suddenly, an opportunity arose.

They were aware of larger "things," things like the host. They knew, from their host, how delicious the big things were, how delightfully emotive their suffering while being fed upon. But, until recently, they'd been too big to prey upon. And they were never alone.

But now? Now, *they* were bigger. Now, the delicious things were dispersing. Now, perhaps, they would be vulnerable.

The pack moved forward in the darkness, approaching the barrier separating them from the big things, reaching out with all their senses.

Yes, they are dispersing!

One at a time, they moved away in different directions. The pack moved along the barrier, searching for the right opportunity: a smaller individual who was alone.

There!

A smaller one was all alone. It stopped moving, then shifted into that strange quiescent state they'd sometimes observed among their prey. They didn't understand this quiescent state. They experienced nothing like it. But they'd learned about it. Their host spent time in this state while they were growing and emerging. The vermin they'd hunted experienced it too, and they'd learned how to manipulate it, to deepen it.

They positioned themselves at the barrier, as close as they could get. Gradually, firmly, they bombarded the prey, deepening its slumber as they began to climb.

■ ■ ■

Donna slowly woke.

She slept deeply, a better night than she could remember. She must have been even more exhausted than she'd realized! And she felt…strange! Gradually, she realized she was lying on her side, her left side. She tended to sleep on her back, yet here she was, on her left side. The room was dimly lit. She vaguely recalled setting the lights to 25% before flopping into her bunk. She thought she'd activate her entertainment, watch some videos before falling asleep, but she passed out right away.

Something was wrong, though she couldn't quite place what. She shifted her eyes and looked up towards the ceiling. She could see some of the tiles displaced, revealing the dark space beyond. Staring into the unexpected blackness bothered her; she should fix those tiles.

She was groggy. So, so groggy.

Donna turned her head. No, she *tried* to turn her head, but her muscles did not respond. As she continued to wake, she realized she could not move: not her head, not her arms or legs, not even a finger. Experimenting, she realized she could move her eyes and her lips. She could control her head, but only her head. Yet she could feel her chest expanding and relaxing as she breathed. Her head continued to clear, and, as it did, she became aware of pain.

Oh my God!! Oh this HURTS! What is happening to me?!

Sharp pain radiated at the back of her neck, like something had sliced her there. It was excruciating! She could feel *tugging* at the back of her neck. Something was crawling on the back of her neck, but she couldn't do anything about it. She couldn't turn her head, she couldn't swat it away.

Her hand! Her right hand was *burning*!

Donna shifted her eyes to look at her hand. Something was engulfing her hand! A weird, alien *thing* had her hand, up to her wrist, in its beaked mouth. Multiple strange eyes on weird stalks waved around, some looking at her forearm, others at her face. She could see and feel sharp claws grasp her forearm, holding it as tongues extended from that beak. Long, red, moist tongues stretched smoothly across her skin, then pressed against her flesh. Sharp needles pressed into her and slowly pulled her hand further into its mouth.

Her head continued to clear, and she felt her hand *burn*. It was on fire! She began to realize that it was *digesting* her hand, working its way further up her arm, feeding on her as she watched.

She felt something else at her belly. She tore her eyes from her hand to look to her belly, barely able to see given the position of her head. She could see another of the things pinching her midsection with its claws. Its beak opened and its tongues stretched into a hole torn in her belly. Nauseated, she realized the tongues were sifting through her viscera, pulling morsels from deep inside her into its beak as it consumed her, tearing away little bits of her and eating them.

Dear God, it was *eating* her!

Then Donna realized her feet burned like her hands, though she could not, no matter how hard she tried, move her eyes in a way that would let her see her legs. She continued to wake, and as she woke, the pain intensified. Rips and burns radiated all over her body. She hurt so much she couldn't think, couldn't stop shifting her eyes from the horror tearing into her gut to the one that was staring into her eyes as it slowly, steadily, consumed her arm.

▪ ▪ ▪

The alpha tore into the back of the prey's neck, exposed its nerve cord, and sifted through the fibers to sever the ones managing voluntary motion while preserving the ones that kept the prey alive and in pain. It no longer suppressed the prey. The alpha felt the prey leave that quiescent state; then, shifting to augmentation, the alpha felt wave after wave of pain and fear from the prey. There were no sounds, though; that was disappointing. The host had made such delightful sounds while they fed. It could also feel its siblings' pleasure as they fed, as their hunger pangs subsided, as they felt their bodies growing.

This prey was perfect! It was smaller. It was delicious. It had been so, so easy to suppress.

Sadly, the alpha realized, at the rate they were feeding, this prey would not live long. Disappointing, but growth and breeding were more important than pleasure. And there were more of this prey to feast upon. At the rate they were growing, they wouldn't need to work together much longer. Soon, they'd each be able to take these prey by themselves.

The alpha shifted its attention from the prey's neck, looking hungrily at the body, deciding where to feast first. It would save the head for last.

DAY 12
JEALOUSY

"How's it going?"

Katya, startled, looked up as Sri entered the conference room. Alone, she'd been totally focused on a wireframe display above the conference table.

Laughing, Sri apologized, then asked, "Is that the residential ring?"

Katya stretched and yawned. "Yes. I was checking on Kiran and Scotty."

Sri leaned in. The display showed two red dots in the interstitial space, presumably Kiran and Scotty looking for their unwanted guests. They were gradually moving away from the maintenance entry portal through which they'd entered the space, their paths diverging.

He spread his hands over the display and brought them together, bringing the rest of the residential ring and the command ring into view. More red dots appeared on the other side of the entry portal, presumably more of Cleo's team preparing to join Kiran and Scotty. He could see a couple of red dots in the captain's conference room: himself and Katya, obviously. More dots in the fabrication center would be Morgan and her team, busily producing weapons for the rest of Cleo's team. A red dot in Sick Bay meant someone was on duty.

He shifted his attention back to the residential ring. Heinz and his team were in their quarters, Amar as well. Simon was in his quarters. The dot in sick bay must be Yvette or Jim.

Reluctantly, he looked at Emil's quarters. Two dots. Emil had company, so it was probably Jim that was on duty.

Sri surprised Katya by swearing, then abruptly pulled his hands together, zooming in on Kiran and Scotty. Without another word, he left the room. Katya sank into her seat, shaking her head, struggling with a strange blend of empathy and resentment.

Sri's interest in Yvette was obvious, as was Emil's. And Yvette's enjoyment of their interest was also obvious. Any fool could see she revelled in their rivalry, but Katya was confident neither Sri nor Emil could see it. No more than she could when Ravi had played such games with her.

Part of her enjoyed seeing a woman playing such games, but a bigger part of her envied and resented Yvette for being able to play such games when she simply could not. She nodded, slapped the table, and sighed.

None of my business!

DAY 12

TENSION

Dee looked over her shoulder. Morgan was busy helping Dawn. The damned fabricator was jammed again. Seemed it didn't matter which fabricator she was using. Dawn always seemed to need help with something!

She leaned over and whispered to AJ, "What the fuck is going on?"

He snorted, looked over his shoulder at Morgan, and shook his head. "Hell if I know! All I know is there's some kind of bugger wandering around. Something from Eden!"

AJ looked over his shoulder again, then checked the readings on his fabricator before answering. He didn't want to fall behind quota. "Eden? What the hell kind of sense does that make? How the hell would something from Eden get aboard?"

Dee checked her readings and said, "right? I don't get it. And what the fuck happened to Bob?"

"He called in sick. Nobody saw him for days. Then they say he's dead, and they won't say shit about how! Something stinks. I've made my share of crossings, and I've never seen anything like this. This ship is *fucked*!"

Dee winced at how loudly he was talking and waved a hand to shush him. "Now don't start talking like that!" She looked around to see if anyone was listening. Morgan was still working with Dawn. She turned back and continued, "You're gonna get us reported! I mean, yeah, I agree, something ain't right, but don't make waves!"

AJ turned to meet her eyes. "Something ain't right on this ship. I've never seen shit like this before."

He glanced around the room before he said, "I'm fucking scared..."

With that, he turned his full attention to his fabricator and refused to answer anything else Dee asked.

DAY 12

LEARNING CURVE . .

The pack fed selectively, as carefully as they could; but, inevitably, the prey lost its grip on life. The alpha sensed its life signs were weakening. It also felt a difference in the quality of the fear and pain coming from the prey.

It remembered this happened with the host. At some point, as the pain and suffering reached a crescendo, the intensity of the prey's experience subsided. The body still experienced pain, but it wasn't processed the same way. The prey's anxiety faded, and that degraded the experience.

The alpha lacked any capacity to understand madness. It only observed the phenomenon. It didn't happen with the crunchy, shelled prey: the roaches and the crickets. So, it had almost forgotten until the warm blooded, hairy prey: the rodents and the cats.

The alpha found this fascinating. It experimented with its prey, trying to bring them as close to the line as it could without crossing it. Just before the line, the waves of pain and fear were intoxicating. But, once across the line, though the fear remained, it just wasn't as satisfying. It had happened to the host, and now, to this prey.

The alpha moved from the meat it had been enjoying where the hind-limbs joined the body to the now lifeless head; it looked to feast on the rich, specialized tissues of the sensory organs and brain. Soon, there was little left. The smaller ones held the larger bones with their claws and used their tongues to get the last bits of nutrition, while the alpha reached out with its senses to find more prey.

There were a number of individuals nearby, but they were big. And, in one case, there were two together, in the same compartment.

There!

Not too far away, past the larger individuals and the two, was another smaller one. The alpha looked over its siblings. They'd grown considerably feasting on this prey. They would work together on the next one, then they should be able to hunt independently.

That would be good; that would enable them to grow even faster!

DAY 12

MASSACRE

Gabby slept soundly as the alpha led the pack into her quarters.

The alpha was pleased. It let each of its siblings practice suppressing her, and each demonstrated mastery. The prey was oblivious to their presence. As a test, the alpha pulled the prey from her bunk to the floor to facilitate their feeding. Briefly, the alpha considered not severing nerve cords, but it remained concerned that the noises the prey might make, delightful as they may be, might bring others to its rescue. It remembered the pleasure of the sounds the host made as they fed and wanted desperately to hear them again. Then, it decided. Later, as they continued to grow, they'd be able to enjoy the sounds made by their prey again.

Once the proper nerve cords were cut, they let the prey's mind clear, then began augmenting its sensory experience. Those positioned to stare into the prey's face experienced a thrill as its eyes fluttered open, and they all reveled in the fear radiating from the prey. They approached, tearing away the material impeding their access to the prey's flesh. The alpha stared into the prey's eyes as siblings reached out to draw the ends of the prey's limbs into their digestive organs, as others carefully opened its abdomen to selectively rasp morsels from within its body.

The alpha decided to experiment. Staring into the prey's eyes, carefully grasping the prey's head with its claws, it opened its beak, everted its digestive organ, and pressed said organ against the prey's face, slowly releasing digestive juices, absorbing the prey's tissues as they liquified. As it did so, the prey radiated fear and agony. The alpha and its siblings wallowed in the pleasure of these sensations as they enjoyed the flavors, working slowly to maximize the experience, keeping the prey alive as long as possible.

When the prey finally succumbed, the alpha tensed its claws to crush the skull. They consumed all that remained, digesting all but the largest skeletal structures. They'd grown so much, but the alpha remembered all too well the smell of the alarm pheromones released when others were brutally killed.

The alpha decided it was time.

Without warning, it snapped out its claws, grasped the smallest of the seven, and used its beak to snip off its sensory cluster. It quickly dismembered its sibling, letting each of the surviving six consume its share. When they were done, the alpha released sex pheromones and was pleased when its siblings did as well. It was intoxicating. They each felt a stirring in their backs as their sex organs began to unfold, stretching from their protective sheaths.

They lined up in pairs, staring into each other's eyes as they positioned themselves front to back so that each could reach, with its male organ, its partner's female organ. As cups covered tips, as spines pierced flesh, as fluids pulsed between their bodies, they trembled with exquisite pleasure.

At some level, they knew not all was as it should be. That there should be more to the process. But there was no time. There was a limited amount of prey, and something, perhaps the larger prey, had been brutally killing their kind. Nothing was more pressing now than mating. Then they had to resume feeding, in order to sustain the development of their offspring.

▪ ▪ ▪

Simon could not recall being more exhausted, not even during his residency. He passed out as soon as he lay down, slept soundly until his bladder could no longer be ignored, then stumbled back to his bunk.

He sat, intending to lay back down, then froze.

Unable to move, unable to utter a sound, Simon stared into the darkness as something lowered itself into his quarters. It moved towards him and, as it did so, entered the dim lighting from the water closet.

It was like the creature he'd found in Esposito's quarters! The creature he'd captured, then placed in that enclosure. But it was *huge*! It was bigger than the biggest dog he'd ever seen, almost as big as he was! He stared, transfixed, as it approached. He saw colors flashing along its flanks as the claws reached out for him. Frozen by fear, manipulated by signals he could not understand, unable to move, unable to call out, the claws gently, then firmly, clasped his arms, pulled him forward, and drew him down onto the floor.

One of the claws released its grip as the other turned him onto his back, then let go as well. He felt his clothes rip off his body, then felt the claws take hold of his torso as warm, moist, muscular tongues pulled his feet. They were smooth as they stretched across him, then painfully abrasive as they took hold of his calves and tore at his flesh.

Simon felt something wet and smooth pressing against his feet; then, as the tongues and claws tightened their grip, he felt his feet begin to burn. He stared upwards and saw the creature's eyes, weird and alien, staring into his own.

Suddenly, Simon realized he was being eaten alive, and began to scream.

▪ ▪ ▪

Amar groggily realized he was hearing screams.

What the hell? Who is screaming? And at this hour?

He tried to move. In his mind, he pressed his hand against the surface of his bunk to lift himself into a sitting position. Yet, strangely, he wasn't moving.

Fascinating!

He realized he couldn't move anything but his eyes. But there was nothing to see. His room was completely dark.

Odd!

He couldn't see, but he could *hear* something. Part of his mind was amused by the irony of his obsession with sight, especially given the darkness. He'd come across some interesting research a while back about human prioritization of their visual senses, about apes in general, really. It reflected in the relative development of the parts of the brain supporting vision compared to our other senses. Not a particularly useful study, in his opinion, especially given his focus (yet another part of his brain noted the unintentional pun) on the highly abstracted mathematics required for astrogation and achieving relativistic speeds which, as one neared the speed of light precluded one's ability to actually process visual input.

Another part of his mind focused (there's that word again!) on the irony that both the darkness and his inability to move precluded his ability to exploit his primary, preferred sensory modality; but, neither darkness nor immobility prevented him from *hearing* the sound of something large moving in the darkness. He could hear the scraping of a chair across the floor as something pushed it aside or bumped into it or something. But whatever it was wasn't making any noises of its own, not at all.

And he could smell something, something not familiar at all. Yet another part of his mind began thinking about the irony of his ability to multitask. Nurtured since his youth, as he trained to become an astrogator and manage the complex multidimensional mathematics of his chosen profession. He was somehow processing all of this yet studiously ignoring the entirely understandable terror he *should* be experiencing under the circumstances...

Suddenly, he was moving. Or, rather, and to be precise, he was being moved. Now *that* suddenly demanded his full and complete attention!

Amar felt the pressure of being gripped by something (his mind raced through possible visuals based on the physical sensations he was experiencing). Then, he felt himself being lifted slightly while being dragged across the bed to pull him from his sleeping position to the floor. Then he was released. Then he felt smooth, moist flesh slithering against his own skin, then felt thousands of sharp needles pressing and pulling against him as his clothes were torn away.

Well, this is certainly peculiar...

Suddenly Amar was overwhelmed with sharp, intense pain. Something tore into his belly! He was held in place, and something was ripping into his abdomen. The relentless analyses of his mind were abandoned completely and his universe became a long, singular experience of agony as something pulled him apart and sifted through his viscera.

He could not see.

He could not speak.

He could not scream.

He could do nothing as it went on and on and on...

▪ ▪ ▪

Emil finished relieving himself and turned back towards the sleeping area.

He'd set the water closet lighting to 20%, so as to not disturb Yvette. But it was enough light to admire her as she slept. He shook his head, knowing this was getting awkward. For DSS lifers, career officers and crew, it was normal and healthy to satisfy each other's needs. It was at least understood if not expected.

Pairing off was challenging, especially for a crossing with only twenty-five souls aboard.

Emil first met Yvette during the crossing to Eden. They each continued to see other people during that crossing, though the frequency of their evenings together gradually increased. When he learned that she planned to join the return crossing, he was surprised by his own relief that they'd be able to continue seeing each other.

But somehow, with the smaller crew, he found himself experiencing jealousy. Somehow, the limited number of alternative partners she might choose made things worse. Sure, he spent the occasional night with Katya and Morgan, but, as far as he knew, Yvette's only other partner was Sri.

While he had not broached the subject with Sri, he was sure this was the source for the growing tension between them. This wasn't healthy and would need to be addressed. And soon, as soon as they got rid of these things from Eden. Should he broach this with Yvette? Should he tell her how he was feeling? Exclusivity under these circumstances would be awkward, but not nearly as awkward as having his suggestion of exclusivity rebuffed. What then? Would she, effectively, be exclusive with Sri to avoid deepening Emil's feelings?

Shit...

Emil was a deep space lifer. He'd made many crossings; he'd been with a lot of partners. He'd never felt any desire for monogamy before. But there was something about this woman that inspired thoughts of settling down, starting a family.

Was it her? Or was it him? Was something changing in him that brought these thoughts to mind?

He shook his head again, then looked at her as she rolled onto her back. Something about that position moved him, and he felt a lustful stirring. He started moving towards the bed, then realized he wasn't moving.

Odd...

He looked back and realized she was moving. Somehow, she was sliding, feet first, across the bed. He looked harder, and he realized something was there. Something was pulling her across the bed! It was one of those things! One of the monsters they found in Esposito's quarters! It was *huge*!

Emil stared on in horror, unable to move.

He watched it pull her feet into its beak. He watched as she appeared to waken; she lifted her head slightly, as the thing moved its eyes across her body to stare into her face. He tried desperately to move, to call out, to do *something* to help.

Then he realized he was being pulled aside.

Weird, strange eyes looked into his own. Emil felt his back press against the wall of his quarters, felt something manipulate his limbs. His legs pressed together, his arms glued to his sides. Something moist and grey emerged from the beak of the thing in front of him, and pressed against his body.

As the pressure immobilized him, Emil's mind cleared. He tore his eyes away from the thing before him and looked at Yvette, as the creature stretched its tongues across her naked body to pull her deeper into its mouth. Her hips pulled in as she arched her back and moaned in agony. Then his eyes snapped back to stare at the creature holding him, pressing him against the wall, folding his limbs against his body as he was smothered in a warm, moist, grey muscle.

When Emil was finally fully immobilized, as he stared into those awful alien eyes, he screamed as he felt the digestive juices begin to burn.

DAY 13
DESPERATION ..

Katya returned to the projection. She'd permitted herself an hour's nap on one of the cots, but she was still groggy. She knew sleep deprivation was likely affecting her thinking.

Can't be helped...

Heinz's team should wake soon. They'd need a little time have breakfast before relieving Cleo's team. They just finished fabricating torches for Morgan's team and would take the next rest break. Heinz's team would produce weapons for themselves, then relieve Morgan's team, and they'd get some level of sanity back in their lives.

Red dots representing Kiran and Scotty, now joined by Cleo, Nils, Giri, Eddie, and Lisa, were spread across the residential ring's interstitial space. So far, they'd found nothing. She was briefly distracted when Sri returned to the room and took a seat across the table from her own. He looked at her and asked, "any signs?"

She shook her head. Sri shrugged, then walked towards the counter to fix himself something to eat. As he did so, Katya leaned back.

"Why isn't Kiran moving?"

Sri returned and looked at the display. She was right. Kiran wasn't moving.

He tapped his wrist computer, then asked, "Kiran, why aren't you moving?"

Silence. He tapped again.

"Kiran? Kiran, why aren't you moving? Do you copy?"

He tapped again and said, "Eddie? I can't raise Kiran and he's not moving. Please go see what's going on with him, ok?"

"Copy."

As they watched, Eddie's red dot moved towards Kiran's as Katya's belly filled with dread.

...

As he walked, Eddie tapped his wrist computer to increase the range and breadth of his headlamp's illumination. He'd briefly considered asking Katya to simply turn on the lights, but knew they didn't want to startle their unwanted guests. It was silly, really. The sniffer wasn't indicating any of the little bastards were in the area anyway, so why not turn on the lights?

Eddie grunted. In truth, he didn't care for Kiran. Guy had a weird sense of humor, a bit of a sick bastard, really. Maybe he'd gotten bored and come up with some stunt. Never knew with that guy.

Eddie kept walking. *Should be another 20 meters or so…*

"Ping!"

He stopped and looked at the sniffer affixed to his flamethrower. Yep, it was picking up one of the little buggers. Must be fairly close if the sniffer was picking up enough for the thing to ping. He swept it slowly side to side. It pinged again, in the direction he was already heading. Kiran must have found one.

Probably played with it instead of just incinerating it.

He'd found Kiran tormenting a roach once, tearing the poor thing's legs off. Nasty little things to be killed on sight, sure, but it should at least be a humane death. Same with their unwanted guests. Just incinerate the damn things!

Eddie resumed walking. He had to admit, he was curious about the things. He'd seen the one in the captain's conference room. Weird little bugger. Amazing camouflage! Hard to believe the things ate Esposito!

Poor Bob must've been really sick.

He paused and shook his head. His mind was wandering, and he realized he'd veered a little to the left. His vision was getting fuzzy, too. Man, he was tired! He wished it'd been their team's turn to get some sleep!

Eddie stumbled, then recovered, grinning sheepishly. He must be more tired than he realized! He took a few more steps, then slowed.

There was something on the floor.

He had difficulty making it out. He walked slowly forward, squinting against the fuzziness; he shook his head, swearing under his breath. He stopped and crouched. The shape was strangely familiar, but he couldn't quite process it.

It was a bone.

He stared, trying to make sense of what he was seeing. Eddie turned his head to the right, then left. In the illumination of his head lamp, he saw shredded, bloody clothing, more bone fragments, a few scattered teeth.

A crushed skull. He was standing over a crushed, *human* skull.

He shook his head again, then gasped as he processed what he was seeing was, in all likelihood, Kiran.

What the fuck happened to Kiran?

A noise…

He turned slowly, too slowly, towards the noise. It was one of the things! It was *so big*!

He stared, unable to move as it approached him. Weird eyes moved as it stared at him. Bright colors flashed along its flanks as claws reached out for him.

He was so tired. So, so tired…

Eddie felt his body fall as his legs gave way; he felt wetness as he lost control of his bladder. He fell to his side and felt such relief as claws closed around his chest and waist, taking his weight for him, letting him rest. The grip of the claws was painful yet pleasing. There were sharp edges and spikes pressing into his flesh, but he didn't mind. He felt loved, supported, cared for.

Eddie smiled as he stared into two alien eyes as the others scanned his body; he smiled as the beak opened and long, red tongues reached out and tore away his clothing. He continued to stare into the eyes as the tongues explored his body. Smooth, moist, muscular tongues slid over his skin, then tensed and began to tear away at his flesh; it was painful, yet so pleasant, almost orgasmic.

Eddie felt a beak pull in his left leg, and watched it slowly, deliberately close, severing his thigh. The pain was incredible, but it felt so, so good, and it was just so nice watching the thing swallow his leg. He felt it shift the grip of its claws as it slowly, methodically bit off his limbs. He grinned happily as it used its tongues to pull his hips into its beak. He felt warm moisture as he was pressed deeper and deeper into its gullet, a growing, awful burning sensation as it began digesting him.

Slowly, as the tongues pulled him deeper and deeper, Eddie's mind began to clear; his brain began to process that he was being eaten, eaten alive by this horrible alien nightmare. The tongues caressed his shoulders, his head, and neck. His body was on fire, and Eddie began to scream.

■ ■ ■

Katya leaned forward.

"Eddie, what's happening? Did you find Kiran?"

Nothing.

She turned to Sri. "I have a bad feeling about this…"

A moment later, she heard Cleo's voice.

"Katya, I can't be sure, but I think I just heard a scream. What's going on?"

Sri stood abruptly, scanning the wireframe display.

"I want you guys together. Uh, Cleo, you stay put. Everybody converges on Cleo's position. Tell me when you're together!"

Sri turned to Katya. "I want the lights turned on!"

She nodded wide-eyed with fear, then tapped her wrist computer.

▪ ▪ ▪

Cleo was surprised when the lights snapped on. The space hadn't been illuminated since the reconfig. Weren't the lights kept off so it didn't want to spook the pests? She turned, slowly, looking around. She could see Nils and Giri walking towards her. Behind them were Scotty and Lisa.

As they approached, she heard Katya say, "Cleo, I want you to walk, carefully and together, to Kiran and Eddie's position. They're not moving, and they're not answering. I want you to be careful, you hear me?"

They looked at each other a moment before Cleo acknowledged Katya.

Lisa wagged her head in disbelief. "What the fuck?"

Cleo stared at the floor, numb to the situation. "No idea. Let's go."

▪ ▪ ▪

Two creatures, startled by the sudden brightness, detected the approach of the others. They froze, matching the colors and textures of their background. It was challenging. The area was mostly flat, open, and exposed. But there were a limited number of vertical structures, and positioning themselves near them seemed to help.

They'd fed gloriously, albeit too quickly. There were more prey in the area. They'd grown, certainly, and they'd learned so, so much about manipulating these delicious creatures; but the memory of the alarm pheromones had not yet faded. They needed to be careful; and, as disappointing as it was to deprive themselves and their prey of the full, glorious experience of their suffering and consumption, they needed to be ready to feed again.

Or defend themselves…

▪ ▪ ▪

Cleo was a couple of paces ahead, Lisa to her left and Giri to her right. Nils and Scotty ran a few paces behind.

They held their weapons ready, heads on swivels, scanning the area. Cleo occasionally glanced down at her sniffer, which was beginning to register the presence of their unwanted guests. Not enough to elicit a "*ping*," but it was there, and it was growing.

She paused, then crouched a tad and squinted. There was something on the floor. She raised a hand to get the others' attention, then walked towards the clutter. Bending over, she pulled a pen from the shoulder pocket of her coveralls. Poking at the clutter as the others joined him, she realized what she'd found.

She stood abruptly with a hiss. "I think it was a cat!"

Giri crouched to take a closer look, then nodded.

"We gotta report this!"

Nils nodded.

"Katya? Katya, we've found what appears to be skeletal fragments from a cat. I think our guests ate a cat!"

▪ ▪ ▪

Sri leaned back, then looked at Katya.

"How the fuck would these things eat a cat?"

Katya stared at him, stupefied.

"They ate Bob!"

Sri, skeptical, responded. "Nils? Proceed. Cautiously! But proceed. Stay in contact!"

▪ ▪ ▪

Cleo exchanged glances with Giri and Lisa, and continued moving towards where Eddie and Kiran should be.

"*Ping!*"

She stopped and looked at the sniffer, realizing there must be one or more of the things in the area. If they were able to take a cat, did that mean they'd grown? How? The ones the gizmos found in the ductwork hadn't grown that much! But if they hadn't grown, then how'd they take a cat?

She looked at Lisa, Giri, and Scotty again, then continued walking.

But after a few more meters, she felt...*strange.*

▪ ▪ ▪

Katya, sitting in Sri's chair, chewed her thumbnail. Sri paced around the table. Over the table, the wireframe showed seven red dots clustered together. Seven red dots, Cleo and her entire team. They weren't moving.

Why aren't they moving?

Sri stopped pacing, walked to the table, and, staring at the display, asked, "Cleo? Cleo, what's happening?"

Silence.

Katya whispered in fear, "I don't like this."

Sri shakes his head in disbelief. "We'd better wake everyone up. Would you do that, please?"

Katya moved away from the table to call Emil as Sri called Morgan. After a moment, he heard Morgan's voice and a lot of background noise.

"Let me step out of the fabrication center."

A few moments later, Morgan asked, "What's going on?"

As Sri leaned forward to answer, Katya came back. "Nobody's answering!"

Sri, feeling a chill of dread, opened his mouth but was interrupted by Morgan.

"Who's not answering? What's going on?"

Sri slowly stood, then held his hands over the wireframe and moved them together. As he did so, the wireframe display zoomed out to show the command and residential rings, with a red dot for every soul on board. Well, a red dot for every wrist computer on board…

His eyes darted about the display, noting the highlighted persons: a cluster of randomly moving dots in the fabrication center for Morgan and her team, two moving dots here in the captain's conference room for Katya and himself, and a cluster of immobile dots in the interstitial space and more immobile dots in the officer and crew quarters in the residential ring. He instructed the display to suspend the display of anyone that had not moved at least a centimeter for at least an hour. All the dots disappeared except for himself, Katya, Morgan, and her team in the fabrication room.

"Holy shit…"

"Captain? What the fuck is going on?"

Sri took a deep, trembling breath. "We don't know what's going on. We lost contact with Cleo's team in the interstitial space. None of the off-duty personnel are responding."

"What do you mean 'not responding'?"

"I mean no one is answering us, and sensors say they haven't moved in at least an hour. I mean, even if they're asleep, they'd be moving, right? Shifting position? And wouldn't you expect somebody to get up to take a leak?"

After a few moments, Morgan asked, "what do you want me to do?"

Staring at the wireframe, Sri straightened his back. "I want you to arm yourself and your team. I want you to send four team members, in pairs, to do a well check on the off-duty personnel. I'll do an emergency security protocol override to unlock everyone's quarters. That leaves you taking someone with you into the interstitial space to see what happened to Cleo's team."

Another pause.

"Acknowledged."

■ ■ ■

Morgan stepped back into the fabrication center, beckoned her team to join her in the corridor, and explained what she'd just heard. She gave them a moment to process what they'd just been told, then directed Ella and Dee to check on Heinz and his team, and AJ and Dawn to check on everyone else. Turning to Izzy, she said, "That leaves you with me in the interstitial space."

■ ■ ■

AJ and Dawn stepped cautiously into the captain's quarters, raised the lights to 100%, then froze.

On the floor, by the wall, shredded, bloody clothing and a wrist computer glistening damply. Above them, the wall wore a roughly oval stain, bigger than a man. AJ stepped towards it and wrinkled his nose as he picked up a foul, acidic smell from the stain. He crouched and, using a relatively clean bit of cloth, tapped the wrist computer. It was Emil's.

He turned towards Dawn, then noticed she was staring at similar debris on the floor at the foot of the bed. The sheets were bunched, suggesting something, or someone, had been dragged from the bed to the floor. He heard her mutter, "What the hell happened here?"

He shook his head slowly, then scanned the room, stopping when he saw a fair-sized lump. Well, two lumps, one dark brown and the other white. Something about them bothered him, then it clicked.

It looked like the turd Simon's pet left in that cage, but *bigger*. Much, much bigger. He cocked his head, but kept his eyes on the dropping.

"Do you think that's what I think it is?"

She didn't answer at first. She was staring at her wrist computer, then held out her arm. She turned, swinging her arm in a broad arc. When her computer made a pinging noise, she stopped and reversed the swing until it pinged again. She repeated this motion, each swing smaller, until she was pointing at the dropping with her computer pinging steadily. She turned to look, then her eyebrows raised with recognition.

"Holy shit! Yvette's wrist computer is *in that shit*!! Fuck me!"

He nodded his head slowly. "Captain?"

▪ ▪ ▪

Sri stopped pacing and answered. "Yes?"

"We're in Emil's quarters. We found…well, it's not quite clear what we've found!"

He described the scene as best he could, then added, "We checked their computers. Emil and Yvette. We, uh, found Yvette's computer in what appears to be a huge dropping. And there's no sign of either of them."

Sri felt a pang, then a chill in his gut.

"What do you mean, there's no sign of them?"

"I mean…there's no sign of them! There's stuff on the floor, like they just, suddenly, disappeared. There's a stain on the floor by the bed and another on the wall. The stains smell…acidic? But nothing else. No sign of them, no trail suggesting they were dragged away."

Sri exchanged a look with Katya. They could hear the fear in Dawn's voice and AJ muttering obscenities in the background.

"Dawn, speculate. What do you think you've found?"

"We think they're dead. We think the critters have grown. A lot. We think they killed Emil and Yvette, and we think they were consumed, right here. Don't know what else to think, dammit!"

Katya sank into a chair, a look of abject horror on her face.

Sri, steeling himself, said, "Dawn, AJ, I need you to check on the others."

He broke the link before they could respond, and said to Katya, "You'd better raise Morgan and see what she's found."

▪ ▪ ▪

Morgan crouched to look over debris as Izzy scanned the area. She stood and asked, "What do you make of that?"

Izzy glanced down at the hodgepodge of torn, bloody clothing, wrist computers, portable incinerators, and other personal paraphernalia mixed with sticky residue strewn across the floor.

"Same as you. That's blood, blood and God knows what else. And I don't see any sign that anyone or anything was dragged away."

Glancing nervously around, Morgan asked, "What the hell could do that?"

Izzy snorted. "I don't think there's much doubt, is there? Gotta be those things. Fuckers ate Bob, right?"

"I know, but they left bones and shit behind, right? Not like this. There's… nothing!"

Izzy swore. "What else could it be? What the fuck else could it be? People don't just…disappear!"

Morgan nodded, then crouched to check an incinerator. She stood and walked to the next one and checked it, too. After she'd checked five of them, she reported, "They're full. No sign any have been fired. How could they be attacked without even trying to defend themselves?"

Izzy didn't respond, so she turned towards him. He was just standing there, immobile, staring. She continued to turn to see whatever it was he was staring at.

As she turned, she felt…odd.

▪ ▪ ▪

"Morgan? Morgan, please respond!"

Silence.

Katya sat, elbows on table, hands on her face, muttering, "oh, my God, oh, my God. This can't be happening!"

Sri leaned back. "They're not responding! Cleo's team is gone, and now Morgan and Izzy are not responding!!!"

"Katya? This is Ella. Heinz' quarters are empty. His wrist computer's here, and there's some shredded clothing. That's all!"

Katya stared at the display.

"Captain? Katya? Anyone?"

Sri hesitated, then answered, "yes, Ella. Yes, we hear you. Ella, please be careful. Please continue to check."

Katya stared at Sri. "Are you nuts? They need to get back here. The four of them need to get together and then they need to get back here! Right away!"

Sri started to respond, then snapped his eyes back on the display when they heard a gasp. He realized the link with Ella was still live.

"What the fuck is that?"

"Ella? Ella, snap out of it!"

"Ella?"

Katya yelled, "Dee? Dee, what's happening?"

There was a pause, then Dee responded softly. "I…I don't know, exactly. Ella walked a few meters ahead of me, and now she's standing in front of Mike's quarters, and she's suddenly not moving. Hang on, I'm catching up with her."

"NO!! Dee, NO! Stop walking! Stand still!!"

"What? Why?"

"I…don't know. A feeling. I don't know, but I'm wondering whether these things can hypnotize you or something!"

Silence.

"Okay, I don't know what that means, but she's not moving, and I don't like it. I got my weapon ready, and I'm going to…"

Silence.

Katya yelled frantically, "What's happening???"

"There's something there!" Dee whispered. "Something is reaching out towards Ella! It's like claws or something! She's just…standing there! She's just standing there, letting it happen!"

Sri and Katya could hear Dee's frantic breathing.

"What's happening, dammit?"

Her breathing calmed, then quieted.

"Dee! Dee, what's happening, dammit! Dee, are you there?"

Quiet, even breathing.

Sri looked at Katya, then at the display. Ella and Dee weren't moving.

They heard a soft voice cry out, "no, no, please, no…"

The breathing accelerated, then the shrieking began, then the link was broken.

Sri looked at the red dots corresponding to Dawn and AJ, verifying they weren't in the same ring segment as Ella and Dee. From a life support perspective, rings were divided into quarters, and each quarter subdivided into thirds. They were by Mike's quarters, so…

He shook his head, his face white with panic. "Emergency override! Isolate the second quarter of the residential ring!"

As the display reflected the closing of bulkheads and isolation of duct work, Katya asked, warily, "What the fuck are you doing?"

Sri looked at her. His eyes were wide with fright; an almost manic look engulfed his visage. "Ella and Dee are dead. Or worse than dead. I'm getting rid of whatever's attacking our people!"

Katya stared at him in disbelief. "You don't know they're dead!"

"THEY ARE DEAD!"

Sri held Katya's stare and ordered the computer. "Purge residential ring segment 2, subsection 'A'!"

Neither looked at the wireframe display as the airlock opened, exposing that subsection and everyone and everything in it to the vacuum of space.

"NO!!! No, you can't do that!"

"It's done!"

Katya slowly backed away from Sri in disgust and disbelief when AJ's voice came through.

"What just happened? Felt like a purge!"

Sri straightened. "I just purged 2A."

"What? What do you mean you purged 2A? Why the fuck would you do that?"

"They were dead! It was infested with those things, and nobody was left so I purged it."

Katya's fist pounded the table. "You can't know that! You can't know they were dead! We could have saved them! We could have *tried*!!!"

Sri ignored her.

"Look. Look, AJ, Dawn, I…I don't think there's much point to continuing the well checks. They're all gone. I want you to come back here. Okay?"

Another brief period of silence, then, softly, Dawn answered. "Yeah, ok, we hear you. We're on our way."

"Great. Keep the comm link open, ok?"

"Uh, ok. Why?"

Sri heard Katya mutter behind him, "So he can space you."

"What was that?"

Glaring at Katya, Sri growled, "nothing. Just get back here."

He muted their end of the comm link. "You need to collect yourself. It's just the four of us, now. Get over my decision, and pull yourself together!"

Katya's jaw dropped open, then she shut it quickly. A fire smoldered in her eyes. "I need a minute…"

She stood and walked to the counter for a cup of water as Sri listened to AJ and Dawn make their way through the residential ring. He could see their dots moving through the display as they approached the bulkhead separating the residential and command rings. Then the red dots stopped moving.

He unmuted and asked, "AJ? Dawn? Why did you stop?"

Silence. A tingle ran up Sri's spine. Katya set down her drink and ran back to the projection.

"AJ! Dawn! Respond!!"

Silence.

Sri muttered, "oh, no…"

Katya jabbed her finger at Sri. "Don't you do it! Don't you fucking do it!" She went to the doorway and picked up an incinerator. "Sri, you hold on right there. Don't you do a damned thing. I'm going to go check!"

Sri, eyes wide, wailed, "no! You can't! You can't leave me here alone! What if…what if one of those things is waiting! I can't handle the ship alone!"

"*It's your own damned fault*!" Katya raged at Sri as tears streamed down her face from raw emotion and despair. "You blew this, you fucked it up right from the beginning!" She leaned forward, jabbing the air with her finger, spittle flying as she screamed. "So obsessed about Yvette fucking Emil instead of keeping us alive! Yvette! Precious Yvette! Self-absorbed Yvette, no better than Ravi, and you and Emil fawning all over her! Well, now they're both dead and it's your fucking fault! Well, not AJ and Dawn! I'm going. I'm going to save them! You can pick up an incinerator and come with me, or you can wait here. I don't give a fuck. But I'm going to save them, and you damn well better not space us!"

Sri, confused, stared at her. "Who the fuck is Ravi?" Then, blushing with shame, he ran to the door yelling, "You can't save them! You can't!"

She didn't turn. She didn't hesitate. She just kept walking.

Was she right? Was he distracted by his feelings for Yvette? His jealousy? Embarrassed, Sri hesitated, then stepped back and picked up an incinerator. He walked through the portal to follow her when Sri saw Katya standing absolutely still.

Sri stared in disbelief, then horror, as one of the monsters came slowly around the curve of the corridor and casually reached for Katya. He stumbled backwards, his head shaking slowly, side to side, as claws gently yet firmly closed down on Katya's right shoulder and her left hip. The beak opened and long, red tongues reached out to pull the incinerator from one hand, her wrist computer from the other; then, slowly, casually, the tongues tore away her clothing.

As Sri watched, the thing pulled her feet into its beak. Its snakelike tongues stretched to caress her naked body, then tensed as they pulled her in, their tiny hooks pricking her flesh as her legs disappeared into its mouth. It stared into her eyes as it gradually, deliberately consumed her. Then, as her calves slid into its maw, her eyes came into focus. She stared back at the alien. She moaned, louder and louder, then began to shriek.

Sri stopped backing away, threw down the incinerator, and ran through the curved corridor. He slapped his hand on the wall beside the bridge portal, then ran to the captain's console, identified himself, and demanded emergency control of the ship.

First, he isolated the residential ring segment where AJ and Dawn were attacked and purged all three subsections.

Then he isolated and purged the command ring subsection where that nightmare was feeding on Katya.

He stared at the console, not sure what to do next.

He was alone. Completely alone, the sole survivor on a deep space vessel, hurtling through interstellar space, facing almost three years alone before reaching Alexandria.

Was he really alone? He'd purged the infested compartments. Were there more of these monsters? Should he purge the rest of the ship?

Well, why not? He nodded to himself. *Why the fuck not?!*

He felt a weird blend of resignation, abandon, and a chill of euphoria.

He uttered a quick command, and the bridge projected a display of the complete ship. Odds were that, if there were any more of the things, they'd be in the residential and command rings. On the other hand, it was possible that one or two might be roaming the aft ring by the hanger decks or one of the cargo rings. And, obviously, he knew they might be in the interstitial space because that's where Cleo's team was attacked.

So, he wouldn't just purge segments; he'd purge entire rings. Seal off each ring, one at a time, from the life support core, then purge the entire ring.

He'd started with the aft ring. Then, one ring at a time, he'd progress through the cargo pods, then finish with the residential and command ring. That meant donning one of the emergency environmental suits from the bridge locker, but that would be ok. It wouldn't take long.

Consulting the bridge display, Sri could see the purges would involve a complex sequence of steps repeated for each ring and coordinated between rings, so he instructed the ship to prepare a purge sequence program. While that was running, he decided he might as well set up a distress signal and a short message. He fully expected to communicate directly with Alexandria when they got there, but should anything happen to him before then, well…

They'd need to board and find him, right?

"This is Sridharan Rizzo, Captain of the *Lucky Strike*. I'm the sole survivor. Something, somehow, came aboard from Eden colony, and it killed everyone else. If you're receiving this, I've been injured, somehow, and cannot communicate directly. Please help!"

There. Easy enough.

Wouldn't broadcast until they got within range of Alexandria, of course.

He checked on the purge routine. Almost ready to run. Best fetch an environmental suit. He could always record a better message later. Plenty of time later!

He stepped away from the bridge console and walked towards the locker.

The portal! It's still open! Dammit!

He turned towards the portal to lock it but, as he walked, he slowed.

He realized he was feeling a little fuzzy…

PART 2
ALEXANDRIA

DAY 1

15 YEARS LATER. . .

"Guys, we're getting close!"

Miri, at the controls of the shuttle tug, leaned over the console, staring through the panel. At Marcus' direction, she'd plotted an intercept course with the *Lucky Strike* and could not wait to see it. Deep space vessels are *so* huge; seeing one is always a thrill.

Marcus and his team were also thrilled by this opportunity. They were eager to pick up any work they could. Since launching their business, their jobs had been modest: routine repairs, maintenance, recoveries, that sort of thing. They only had the shuttle tug, and its limited speed and range were a constraint compared with more established competitors.

Marcus had a dream, a dream and a plan.

The plan started with enrolling his daughters, Miri and Robin, with the Deep Space Academy. He'd considered a PIP crossing to get them a free education, but realized their mother's passing was too recent, their emotions too raw. Instead, loans allowed Miri to earn her shuttle tug license and Robin her harbor pilot's license. That let them earn a little here and there to help their father repay the loans that funded their education. Marcus, with his buddies Dan, Brad and Reece, hired out to any repair, recovery, or salvage operation they could find while Marcus searched for a shuttle tug that they could afford.

Finally, he found one. It was old and needed considerable work, expensive work. But soon, it was *their* shuttle tug, *their* capital investment, with very steep payments. An expensive foundation for their aspirations, but they had to start somewhere.

It was a big financial risk, but it was *his* risk. Miri and Robin were committed; but as their father, he shielded them from his exposure. As for Reece, Dan, and Brad, well he'd bet his life more than once on their competence and integrity, as one does while working together in space. But they were in no position to help. In space, they were solid. In life? Total screw-ups. Not a lick of financial sense in any of them. Every payday, as soon as they satisfied whatever debts they were carrying, the rest burned holes in their pockets. And, once they'd blown through whatever they had, they started accruing their next round of debts.

Dan and Reece's vices were binge drinking, and Brad's was the red-light district. Marcus wasn't a prude, and he'd certainly done his share of carousing as a young man. But he'd grown up since meeting his late wife and becoming a father. Dan, Reece, and Brad could continue to live in the moment, but now? He carried sole responsibility for Miri and Robin, and he needed to build something.

He sighed. As long as they always showed up for work on time, he didn't feel it was his place to judge. Well, mostly. He wasn't quite sure how he felt about the way Brad sometimes looked at Miri. Might be his imagination, might not. But, as long as Miri didn't seem to be returning the interest, it seemed academic.

He shook his head and peered through the panel. This assignment would change everything! Salvaging a deep space vessel! Intercepting and docking a deep space vessel was a plum assignment. Even a spaceport as advanced and busy as Alexandria only saw a handful each year, and such opportunities usually went to one of their more established competitors.

But not this time.

The *Lucky Strike*'s distress signal was received a few days ago and was accompanied by a brief and vague message from her captain. Something about creatures from the Eden colony killing his officers and crew. Other than that, the *Lucky Strike* was eerily silent, suggesting the captain had succumbed to whatever it was.

The *Lucky Strike* was on autopilot, cruising towards Alexandria at space normal speed. That normally wouldn't be such a problem. Sure, once the ship cruised past the colony, whomever won the contract would have to chase her and bring her back. Unfortunately, whoever put her on autopilot wasn't aware that, since leaving Alexandria some thirty years ago local time, they'd started construction of a second orbiting platform and spaceport. He or she couldn't have known that they'd set a collision course.

There simply wasn't time to wait for one of their more experienced and better equipped competitors. Somebody needed to intercept the deep space vessel ASAP and find a way aboard, which would be challenging with nobody aboard to open a hanger bay door for them. Then, once aboard, someone had to take control of the bridge and shut the ship down so she could be towed to a space buoy.

Marcus and his team were the *only* option.

When Marcus won the assignment, Miri, knowing deep space vessels were normally towed to their buoy by two shuttle tugs, reserved simulator time until she felt confident they could do it with one. There'd been a bit of back and forth about compensation. The usual docking fees were substantial, but the salvage rights were unclear. The immensely valuable cargo was already sold, and the proceeds belonged to Eden colony. On the other hand, the presumably deceased crew had bonuses coming that, it was decided, would be awarded to Marcus and his team. It was Brad that insisted they pool enough of each of their shares to pay off all their loans, which touched Marcus deeply. Of course, that still left each of them with a tidy sum. He helped Miri and Robin manage theirs, making investments in their future but reserving a percentage for splurges. Brad, Reece, and Dan…well, he knew he wouldn't see any of them for a while once they were paid.

Then there were the personal effects of the officers and crew. Given time dilation, deep space officers and crew very rarely had surviving relatives close enough to make an inheritance claim. So, as with the bonuses, personal effects would also go to Marcus and his team. If, of course, they could figure out a way to get Robin aboard the *Lucky Strike*. Worst case, they'd secure the shuttle tug to the deep space vessel's drum and cut their way through.

"There she is!"

Miri pointed to starboard. Everyone secured whatever they were working on and launched themselves towards the cockpit; then, taking advantage of zero-g, Marcus's team arranged themselves over the controls, allowing them to peer out without getting in Miri's way. All were quiet as she manipulated the controls to adjust the attitude of the shuttle tug. Now that they were oriented to approach the *Lucky Strike*, the immense craft filled the portal. No matter how many times one sees a deep space vessel, they're awe inspiring.

After a few minutes, Marcus pointed and said, "When we're closer, I'd like to do a spiral flyby to visually inspect the ship before we board her. Maybe 75 meters?"

Miri nodded. They'd discussed this before heading out. They needed to know the condition of the ship, and they needed to decide their best penetration opportunity. Miri, adjusting to the deep space vessel's velocity, did a quick cruise around her forward funnel. The funnel, designed to shield the drum while collecting interstellar matter as fuel, had about as many dings as expected. The star drive, when engaged, collected material with that forward funnel. To accelerate, the ship passed that material through the main shaft to be annihilated by antimatter in the aft funnel. To decelerate, the annihilation took place within the forward funnel.

Miri flew a spiral around the main shaft, then the drum, which was not rotating. Brad pointed to the command ring and said, "Hey, there's our entry! Looks like they purged!"

"Why the hell would they do that?"

"Who cares…hey, there are purges in the residential ring! There's one subsegment there, and, wow, all three subsegments over there!"

They pondered what this meant as they continued to spiral around the drum, then the ship's shaft and antimatter magnetic bottles, then the aft funnel.

Once the flyby was completed, Marcus said, "Brad's right, that command ring purge is our best bet. Take us back there, Miri."

Robin, peering out, said, "There should be handles on either side of the airlock. If you grab them with the tug's robotic arms. I'll kick over and enter. Then we just have to close the airlock, pressurize the segment, and hike to the bridge."

Dan and Brad exchanged a look, then Brad said, "If you're comfortable with that…"

She laughed. "If I'm comfortable with that? Hell, you're coming with me!"

Marcus pointed at Robin, then Brad. "I want you two in environmental suits, right now." He turned to Dan and Reece. "I want you two in environmental suits as well, ready to assist as needed. Okay?"

Miri flew the shuttle tug along the deep space vessel's drum, orienting the tug so the *Lucky Strike* appeared "above" them. Flying this close gave each of them a new appreciation for just how massive the ship was. As they passed the residential ring, Miri slowed considerably, then turned to port to creep towards the command ring until they approached the open airlock.

Marcus strapped himself into the seat beside Miri's as she cruised along the ship's main shaft towards the command ring. As she neared the command ring, she slowed to a glacial pace. As they approached the air lock, he took control of the robotic arms and took firm grasp of the handles beside the air lock. Miri released the shuttle tug's controls, and Marcus, using his grip on the handles, turned the shuttle tug until their air lock was directly above the *Lucky Strike*'s. In the air lock, Brad worked the control. Once the shuttle tug airlock depressurized, the outer door opened.

"Still up for this?"

Robin laughed, crouched, and kicked off. As she passed between the airlocks, she turned to absorb the impact of her landing with her feet and legs. She watched Brad do the same, albeit not quite as gracefully. She knew that, despite so many hours in zero-g, he really didn't like it. She reported back to Marcus that they'd successfully crossed over as she watched Brad work the deep space vessel's airlock controls.

As the outer door closed, they moved into the segment. Brad pulled himself to another control panel. The inner door of the airlock closed, and the segment began to pressurize. Given the volume, this would take a few minutes. Meanwhile, she looked around.

"Wow," she said, "whatever happened, it happened before they reached midway."

Brad watched the pressure gauge. "What makes you say that?"

"They're still configured for 'down' to be aft."

He glanced around. She was right. Didn't matter now, of course. The ship's velocity was slow but steady, so they were in zero-g. But, at midway, when the autopilot shifted from acceleration to deceleration, there weren't any humans to reconfigure. So, when gravity reversed and the "floor" became the "ceiling," anything that wasn't secured fell, and a bunch of stuff probably broke.

He grunted and returned his attention to the gauge. That meant the segment had been open to space for at least a year and a half ship's time, maybe more. Maybe a *lot* more. Always a chance there'd be a problem holding pressure.

The gauge turned green and stayed there. After a few minutes, he opened his helmet.

"Breathes good!"

He turned back to the gauges and checked the rest of the rings.

"The residential ring segments that were purged are, of course, still open to space. I don't see a reason to do anything about them, do you?"

Robin shook her head. "No, not really. Are we good to the bridge? I don't want to spend any more time here than we have to..."

Brad nodded and returned his attention to the gauges. After a moment, his brow furrowed as he leaned in towards the display.

"Weird! Some of the command ring segments apparently have slow leaks....Yeah, that includes the bridge. The bridge and the segments to either side of it."

"So, we suit back up? Suit back up and depressurize before going through?"

He tilted his head side to side, then said, "nah. I'm reading the ship's log. They're really slow leaks, but the ship didn't sense anybody in the segments, not anybody alive, anyway, so it closed the bulkheads and let it bleed. Still got plenty of gas, though, so I say we re-pressurize and let it seep until we're done. Might as well be comfortable, right?"

"You're sure?"

He turned and winked. "Oh, yeah, it'll be fine. They're slow leaks, like pin pricks, right? And I don't feel the need to take the time to find and patch 'em. Let the refit guys deal with it!"

Returning his attention to the panel, Brad tapped away. "Okay…overriding the pressure check and…" He tapped again. "Opening the bulkhead. Yeah, the ship's got plenty of air and they're really, really slow leaks. We'll just let 'em bleed air until we're done."

He turned back to the controls. As he pressed the command to open the pressure doors, he mumbled, "Ears might pop if the leaks affect ambient pressure." But their ears didn't pop.

"Must be really, *really* slow leaks!"

Brad wasn't sure what to expect. The air was old, from the ship's compressed supply. Would it be stale? Would there be a smell? Decomposition from the bodies of the crew?

Nothing wafted through the newly reopened corridor.

Robin opened her helmet as well; then, using corridor panel seams, she pulled herself towards the bridge.

"How long do you think?" Brad asked. "I mean, for what it's worth, we could bleed air at this rate for days, but…"

"Yeah, it's creepy. Assuming everything's working, maybe twenty to thirty minutes. I gotta deactivate the autopilot, then hit the retros. It's a massive ship, y'know, in a vacuum. She'll have powerful thrusters, so I gotta be careful to not stress anything while overcoming her forward inertia. Good thing she's not moving that fast! But, yeah, twenty or thirty minutes to ease 'em on and off…"

Robin stopped and stared. Floating in the corridor outside the bridge were shards of clothing and a wrist computer. The clothing, clearly from a deep space uniform, was discolored. She couldn't be sure, but she guessed they were very old blood stains. She turned to Brad, who, seeing her stop, scanned the area until he saw the debris as well.

"Holy shit!"

They both heard Marcus's concern when he asked, "what?!?"

Brad collected himself and replied. "We've found what may be signs that someone died violently. But it looks like it happened a long time ago."

"Should you come back?"

"And then what?" Robin responded. "Let the ship run into that new orbiting platform frame? No, somebody has to go shut down the autopilot so you can tow her to a buoy. We'll stay together and abort at the first sign of trouble, and we'll make it as quick as we can. Okay?"

After a moment, Marcus said, "Yeah, okay. But be careful. No exploring. Get back here ASAP. Once we tow her in, somebody else can board her and figure out what happened!"

"Copy that!"

They made their way towards the bridge. Robin went through the portal to access the command console, and Brad hovered in the corridor. As she worked the console, he drifted slowly past the portal, and, as he did so, saw something around the bend.

"Robin? I'm going to check something out."

Focused on the console, Robin grunted, then responded absentmindedly. "Remember what my dad said! Don't go far, this won't take long, I'm almost done here, I'll be firing the thrusters in a bit..."

Brad turned and reached for the wall to push himself along, then pulled himself through the corridor using wall panel seams. As he progressed through the curved corridor, he could see more and more of the shape, but he couldn't get his mind around what he was seeing.

Suddenly, it clicked.

It was a body.

But not a human body. It wasn't like *anything* he'd ever seen before. It was some weird, alien *thing*. Brad squinted and stared, but he couldn't make sense of it. Didn't help that it was desiccated. He'd seen desiccation like that before. A body, human, vermin, whatever, after a slow leak. No sudden decompression, no boiling of bodily fluids. Just slow, steady sublimation of the body's water as air pressure bled away. Freeze dried, essentially.

He stared a few moments, his mind desperately trying to understand what he was seeing, what sort of creature this once was. But he had no desire to get any closer. Instead, he blindly reached for the wall panel seam, then turned and pulled himself back towards the bridge.

Robin called out, "Brace yourself!"

Brad reached for a wall panel seam as the thrusters engaged, holding fast. After what seemed like forever, the thrusters eased off.

"We've stopped."

Robin came through the portal from the bridge, then stopped when she saw his face. Brad beckoned wordlessly, leading her back towards it. A glimpse was all she needed.

"Let's get the hell out of here!"

She didn't have to say it twice.

DAY 3

TRADITION

As Robin and Brad put away their environmental suits, Marcus used robotic arms to secure their mooring lines. Turning the massive deep space vessel took time and strained their engines; but, slowly, the great ship turned. Miri flew cautiously but confidently, and everything went as she expected it would after all the simulations.

The only unexpected thing was the silence.

Intellectually, as they approached the *Lucky Strike*, they'd known that all were lost. But that was abstract. Reality was a gut punch.

As they approached the buoy, Marcus took control of their tug's robotic arms and, smoothly and confidently, transferred the ship's mooring lines to the buoy. With a nod, Miri flew the shuttle tug back to their space port hanger bay.

Miri broke the silence. "I'm not sure I'm in the mood to celebrate."

After completing any salvage job, especially one this profitable, it was traditional to head to the bar to celebrate. They would buy a few rounds, brag to other patrons, and boisterously compare notes about how they planned to fritter away their shares of the proceeds. After all, even after clearing all their debts, this job left them each plenty to play with.

Marcus grunted, then said, "It's tradition!" With a sigh, he continued, "We'll go to the bar, and we'll raise a toast to their memory. How's that?"

Marcus looked at each of them, and everyone offered a gentle nod. Dan added that he'd go, but he and Reece already booked seats on the next landing craft to the surface, where they planned to drink himself into oblivion. Marcus wasn't surprised. Brad, still visibly shaken by what he'd seen, didn't say anything. He'd also reserved a landing craft seat, and a room in the red-light district, but kept that to himself.

Once they reached the bar, Marcus ordered a first round in memory of the officers and crew of the *Lucky Strike*. They all raised their glasses in toast, then swiftly drank their shots. After a solemn pause, he ordered another round. To his left, Brad lifted his glass, took a sip, and resumed absentmindedly scratching a red spot on his thigh with a tiny black center.

DAY 4

SAD TIDINGS ..

Governor Gulati sat alone in his office, procrastinating.

Yoshi Gulati was a tall, fair skinned, slender fellow with a commanding air. He wore his thick, black hair in a neat ponytail and, even though he'd never served in the DSS, wore one of their working uniforms, albeit without insignia. He found their coveralls a comfortable and practical garment when he didn't need to dress formally; and, since so many of governors *were* from the DSS, well, it fit in more ways than one.

He'd been appointed governor almost eighteen years before, when his boss, Abram Kelty returned to the DSS. He'd taken an entry level position in Kelty's administration, rising quickly to a position with sufficient visibility to earn the Governor's notice. That led to an advisory position to the Governor that became *the* advisory position, making him the Governor's heir apparent.

Despite all of that, he'd never been through anything like the last few days.

An inbound deep space vessel, on autopilot, broadcast a distress signal, bearing a brief and somewhat vague message from her captain, who apparently was the last victim of whatever fate befell the officers and crew of the *Lucky Strike* rather early on in her return journey. A mad scramble ensued to find someone, anyone, who could intercept the ship, get a pilot aboard her, and bring her to a harbor buoy before she cruised past their spaceport and into the frame being built for their second orbiting platform. Once the ship was safely tethered, the salvage team was compensated and admonished to not discuss what they'd found. Not that they'd seen much. All the crew reported was an empty ship, a couple of shredded uniforms, and a very strange, very alien husk.

Yoshi feared speculation would run rampant through the colony, and that was never good.

On the other hand, a number of the harbor personnel and now the salvage team knew all of these things. They all knew they shouldn't discuss this, but human beings being human beings. The odds were high that at least a few friends and family knew something by now. He didn't have much time to get ahead of the rumor mill with real information, so he reached out to his security team and the local deep space academy for help. They kept him posted as they reviewed ship's and officers' logs and sent an engineering team to conduct a physical inspection. When they came across a rather strange carcass, they pulled in some of their xenobiology staff, which of course expanded still further the potential for leaks.

The xenobiologists' necropsy led to a request for a consultation with their Eden counterparts, but he asked them to wait on that. He needed time to digest their reports to date and officially inform Governor Lubandi of Eden.

Which it was now time to do.

He really didn't want to, but it wasn't going to get any easier, and that rumor mill would be running soon if it wasn't already, so he'd best get it over with.

Yoshi stood, walked to his comms panel, cleared his throat, checked his appearance, and started recording.

"Governor Lubandi, I hope this communique finds you well. I know you've been waiting to hear the fate of the *Lucky Strike*. She arrived on schedule, and an automated message was received once she sufficiently decelerated.

I'll forward the message to you, once you've confirmed receipt of this message. It was sent by Captain Rizzo, programmed to be forwarded on their arrival. He reported that a member of his crew was lost to an alien parasitic infestation, and that the parasites, after consuming the remains of their host and exhausting whatever vermin they could find, began to hunt the rest of his crew. He recorded the message when it became clear that it was a matter of time before they all fell prey to the things.

We dispatched a salvage team. The ship was largely intact, and none of the processing pods were affected. Some segments of the command ring were open to space; we believe they attempted to eliminate the predators through vacuum purges. Other segments were compromised by pinholes; they apparently, slowly, lost air."

He paused and collected himself.

"No survivors were found. An alien skeleton, a presumed kraken skeleton based on your reports, was found. The team also found the skeletal remains of what we assume to be the crew member whose initial infection led to the infestation of kraken."

He paused again.

"I'm...so sorry to relay this news. While it is not pleasant news, I hope that it, at least, brings some closure to you and the members of your staff that knew the crew of the *Lucky Strike*. The processing pods will be delivered, and the balance of their sale will be credited to your colony's accounts. The ship will be restored and refitted for her next journey. In addition to Captain Rizzo's final message, we'll be forwarding a complete report of the salvage operation."

He stopped recording, reviewed the message, and decided a second try wasn't going to be any better. He transmitted the message, then returned to his desk to try to deal with various administrative tasks while he waited for an acknowledgment. He didn't often have occasion to exchange messages with Governor Lubandi. When he did, he found her crisp and professional. He wasn't sure how she'd take this news, and he wanted to be available should she have questions. He didn't have to wait long.

He walked back to his comms station to see a projection of Elke Lubandi.

"Governor Gulati, thank you for your message. It is as we feared, and though closure is welcome, the news is still hard to bear. If at all possible, please limit the distribution of this news until I've had a chance to inform those here who knew the officers and crew of the *Lucky Strike* and can make arrangements for a memorial ceremony. Better they hear it from me than through one of the news services. Shouldn't take more than a few days. Thank you."

Relieved that this unpleasant task was over, Yoshi approved initiation of the *Lucky Strike*'s refit for her next journey.

DAY 10
OUTBREAK

Dr. Begay relied on his wrist computer to navigate the primary dome's red-light district. It was in the base station, near the transportation hub.

He was near the end of his shift at the clinic and was looking forward to a quiet evening at home. But, over the last two days, a number of brothels reported providers unable to service clients. Given the nature of their services, managers were required to log health issues. Not that he expected a sexually transmitted disease. Such things happened, but rarely. Still, even routine pathogens tended to pop up here first, especially with deep space vessels in port.

The reporting system evaluating these reports considered severity and frequency. So far, none of the illnesses seemed particularly dire. Providers simply said they were exhausted. Their managers told them to get some rest, logged the incident as required, and went on about their business. But ten reports in the same twenty-four hour period crossed a frequency threshold, suggesting something highly contagious needed to be identified and contained as quickly as possible.

Begay wasn't terribly concerned. He'd reviewed the reports, and it sounded mild. An influenza virus, perhaps, or possibly a coronavirus. Probably carried by a patron from one of the deep space vessels currently in port. But if there had been outbreaks at either vessel's origin, he would have been notified. The absence of such notification suggested whatever this was may have mutated during a crossing. If so, if this disease became more contagious during the crossing, well, once they identified which deep space vessel brought it here, he'd have words with their medical staff about failing to report it! No way they could have something this contagious aboard and not notice!

Dr. Begay sighed as he walked. That was the challenge with influenza and coronavirus, their mutation rate. Thank God whatever this was seemed mild. If they didn't contain it, and it mutated into something more severe, it could really tear through the colony.

Eventually, he found the first establishment.

Should be routine. Take a sample, map the genome, prep a vaccine, and that should be that.

An antiviral would help those already infected accelerate their recovery, and distribution of the vaccine to the rest of the brothels' staff should keep them from catching and spreading whatever it was. Then it would just be a matter of tracking and tracing their recent clients to see whether any of them were infected, who they might have exposed. It'd be a long evening, for sure. The brothel managers wouldn't be happy, but this was standard operating procedure. They understood this would be their least hindering course of action.

As Dr. Begay entered the establishment, he was quickly greeted by an extremely attractive and minimally clad young lady. It took him a moment to regain his composure and explain why he'd walked in. With an abrupt shift in her demeanor, she politely led him past the bar to the manager's office. The manager was a trim, fit, and pleasant older gentleman that stood, shook Dr. Begay's hand, and led him out of the office and down the hallway.

The manager turned while he walked and said, "strangest thing. Maddie told me yesterday she was just too tired to take any clients. Didn't think much of it at the time and let her be. Hey, as long as they make rent for their spaces, right?"

Dr. Begay nodded, and the manager turned back to look where he was going.

"Anyway, their rooms are pretty self-sufficient. Have to be, right? A lot of them just live here for however many days they're on duty. Kitchenette, full bar, lavatory, shower, whatever might be needed, right? So, yeah, we haven't heard a peep from her. Figured she'd let us know if she needed anything, right?"

They reached the door. "Anyway," he turned again to say, "first it was Maddie, then three others through the day. They notified me, and I notified your people, then didn't pay any more attention. But, since you're here, I'm guessing other places are affected, too…"

Begay didn't respond one way or the other. The manager shrugged, then stopped at a door and pressed the call button.

No answer.

He frowned and pressed again.

He turned and said, "Maybe we should just let ourselves in, right?"

Dr. Begay nodded.

The manager pressed his palm on the call button and said, "manager override!"

The portal cleared. The room was dark. As they stepped into the room, Begay heard a quiet moan and an odd, soft, rustling sound.

"Lights, 100%!"

As the lights came on, Dr. Begay could see that they were in a large, predominantly red bedroom. In the middle of the room was a broad, round bed. In the middle of the bed, on her back, moaning softly, lay an emaciated young woman. She was scantily clad, and what clothing she wore was *moving*.

There were *things*, weird, strange creatures, all over her body. One used its claws to pull bits of flesh from her breast. Another used its claws to hold open a ragged hole in her abdomen as its tongues drew morsels from her viscera. A few centimeters to the left, there was another tear through her abdominal wall, and movement under her skin suggesting something gorging on her organs. Another was on her pillow, tearing away parts of her left ear. Yet another, also on her pillow, held her right eyelid open with a claw as dipped its beak into her eye socket and its tongues extended down, pulling away shreds of eye tissue. Another was perched on her thigh, tearing pieces from above her knee. Another had her index finger in its mouth, holding onto it with its claws, the other fingers of that hand naked to the bone.

Begay and the manager stared. The manager blanched, then ran to the sink and vomited. Dr. Begay stood, transfixed, staring at the hell before him, listening to the soft, pathetic moans of what was once a beautiful young woman. Then her remaining eye focused on him, and she pleaded, softly, weakly, hoarsely, in a voice that he knew would haunt him the rest of his days, "...help...me...please...kill...me..."

DAY 11
EMERGENCY MEETING

Yoshi Gulati was not, by choice, an early riser. He was sound asleep when his Chief Medical Officer, Jane Carson, called an emergency meeting. He responded that he'd need thirty minutes to pull himself together, so the meeting wasn't scheduled to start until 06:15. He resisted the urge to complain about the time, because he knew that Jane knew his habits and wouldn't have scheduled something at this ungodly hour if it wasn't, truly, an emergency.

His quarters, offices, and conference room, clustered together, comprised a suite taking up the western side of the top floor of the base station anchoring the space elevator. It was, undeniably, a practical arrangement. And rank hath its privileges. But, it also meant that he spent most of his time in a relatively limited area. The challenge was always balancing a busy schedule against living in a bubble.

Yoshi broke his fast, dressed, then, before walking from his quarters to his conference room, decided to finish his coffee by one of his western facing windows, as he so often did. He had a commanding view of the western side of the dome. Ribs extended from the gondola airlock to the stem wall, supporting clear panels through which he could see the storms raging to the north and south. Due west, along the equator, was relatively calm. Through the haze, he could make out the residential dome. Somewhere beyond that, lost to the atmospheric maelstrom of his home world, the second settlement was under construction.

He sipped his coffee and thought of Eden, as he'd been doing since the arrival of the *Lucky Strike*. He'd been born here and knew nothing different. His world was in the "sweet spot," the right distance from the sun to have liquid water, but for reasons planetologists continued to argue about, not very much of it, at least not compared to Earth, or Eden.

What surface water they had was rich with organic chemicals, but nothing more. Nothing resembling even the simplest cell. The atmosphere was primordial, rich in hydrogen, methane, ammonia, water, but not much oxygen. Storms raged over much of his world, with a narrow band of relative calm along the equator. Since the discovery of Eden, he'd wondered, as he presumed most did, what that must be like. To see gondolas climbing the space elevator through blue skies to the heavens. To see, in the distance, a horizon. A blue sky meeting a brown desert or green forest or, even, perhaps, a blue sea.

Yoshi checked the time: 06:18. He was a bit late. He drained the last of his coffee and walked from his quarters to his conference room to find Jane, Security Chief Pavel Tocci, Engineering Chief Jack Colbert, and Operations Lead Kim Shelley waiting for him. Jane looked like she'd been up all night. He took his seat, then asked her why she'd called the meeting.

She stood and answered. "We have an outbreak in the red-light district of the primary dome."

"What sort of outbreak?"

With a nod, she waved a hand over the conference room table, producing a projected image of Maddie. Ignoring the obvious pain and nauseating expressions, Jane kept command of the room. Adopting a clinical air, she explained the situation:

"We've been receiving reports from brothels under the primary dome that a number of their providers were unable to service clients due to exhaustion. Receiving ten such reports within a twenty-four hour window crossed a threshold, so, yesterday evening, Dr. Jonas Begay was sent to investigate. He expected to find and treat a routine viral outbreak. Influenza, coronavirus, that sort of thing. Instead, he found this unfortunate young woman. He immediately called for help. We dispatched ten clinicians supported by five of Pavel's security personnel."

She nodded towards Pavel, who nodded back in acknowledgment. "The request had come in after he'd finished his day," she continued. "It seemed routine to the night shift manager, so personnel were assigned without notifying him. He'd only learned of the request when he, per his usual morning routine, checked his messages on the way to this meeting."

She waved her hand again, and two more images appeared next to Maddie: a supine skeleton on a bed, the skull on a pillow surrounded by long, curly red hair, and an emaciated woman encrusted with hard, brown, oblong shapes. Jane pressed on as some in the room winced.

"As of this morning, we've found twenty-three affected sex workers in four brothels. All are female. Seventeen are skeletonized. Damnedest thing I've ever seen. There are no signs of struggle. The sheets are smooth, the bones totally cleaned, and, almost without exception, perfectly placed, as if the victim's tissues had simply disappeared. A few, smaller bones, finger and toe bones, were found elsewhere in the rooms. Very little blood. As you can see, the hair on the pillow also looks like the victim's tissues simply… evaporated. We found three more encrusted with these oblong, brown, shelled objects. These individuals were emaciated, no body fat, muscle tissue degraded, conscious but so weakened as to be nearly unable to move. The remaining three were, like the first image I showed you, being actively consumed by these things."

With another wave of her hand, she replaced the images of the victims with a strange, alien creature. It had six legs, two of which were modified as claws, two tails, a beak, and, sprouting from the top of its body just behind the beak, a thin, jointed limb ending in six eyestalks.

"None of us had a clue what these things are, so we took a DNA sample from one of the creatures and, working with someone from the Deep Space Academy xenobiology department, identified it as *Daimonia horridus*, a predator native to Eden. The scientific name translates to 'horrible demon', and, from what I've now read about them, the name fits. They're also commonly called 'kraken', a reference to their multiple tongues that, apparently, reminded someone of the mythological giant octopus or squid."

She looked about the room. "Some of you are probably dredging up memories of the Eden kraken incident some fifteen years ago. Of the first 1,000 colonists, 143 were lost to a kraken outbreak shortly after they reached Eden."

Yoshi and Pavel looked like they'd each seen a ghost. She wasn't quite sure what to make of their expressions, but decided to press on. She replaced the image of the creature with a looped depiction of the kraken lifecycle.

"For the sake of our discussion, let's start here, with this spore. On contact, it infests the host with parasitic worms. Part of this process apparently tailors the parasite's digestive system to its host, even if that host is alien to Eden. Something to do with the fact that life on Eden, unlike Earth, has five distinct evolutionary trees that are as alien to each other as any of them are to Earth."

She could see from their expressions that her audience wasn't quite following.

"Whatever…doesn't matter right now. The worms spread through their host's tissues, selectively feeding in order to keep the host alive as long as possible. At some point, as they mature, they erupt from the host's body to form a chrysalis…" She pointed to the next stage of the lifecycle, depicting a human host encrusted with brown objects like the unfortunate sex worker they'd just seen. "…where they metamorphose, like a caterpillar into a butterfly, into the creatures we found feeding on victims like Maddie. They skeletonize their host, then start hunting other prey. When they mature, they mate and produce live offspring."

She pointed to the next phase of the lifecycle, looking like a slimmer, winged version of the kraken, and continued, "The offspring are called 'dragons', and you can see why. The dragons are also predators that, when they're large enough, release millions of spores that infect hosts and restart the life cycle."

She paused, then said, "So, there are two, distinct forms: the kraken and the dragon. This is called alternating generations, which is actually common on Earth, but not for any of the Earth animals most people are familiar with. It's common among plants, though…"

She shook her head. "Again, it doesn't really matter right now. The important thing for you to understand is that these creatures alternate between two distinct adult life forms, both of which are voracious predators. The worms and kraken haploid generation is the sexual generation. They mature, mate, and produce live offspring. That offspring, the dragon generation is diploid and asexual. When they mature, they do not mate. Instead, they produce spores. Millions of them. The spores, once released, can remain dormant for long periods of time, until they come into contact with a potential host." Leaving the lifecycle display up, she took her seat, sat upright, and set a formal tone.

"As your Chief Medical Officer, I've informed you that we have an infestation of an alien parasite that has taken seventeen lives. Six more are hospitalized, and I doubt any of them will survive. The parasites are from Eden. We've had two deep space vessels in port during the period preceding when these sex workers began reporting that they were too exhausted to entertain clients. One, the *Eureka*, left a number of days ago. The other, the *Lucky Strike*, arrived from Eden almost a week ago. Of the two, the *Lucky Strike* is obviously the more likely source, so we began the process of trying to find the personnel from that ship. But, when I asked the night shift Harbor Master for a personnel list so we could find them, I was told that, even though she's tethered to a buoy, we have no record of any personnel leaving that ship."

She turned to the governor and concluded, "Yoshi, can you help us understand what's going on here?" She let her voice trail off.

Pavel turned to look at the governor as Yoshi leaned back in his chair, closed his eyes, and pinched his nose while the rest of the room looked on. Finally, he released his nose and meshed his fingers as he leaned forward.

"What you don't know," he explained, "and must now be told is that the *Lucky Strike*, which arrived from Eden, arrived on autopilot. I wanted to respect Governor Lubandi's wishes, so I limited the distribution of this information." He nodded towards Security Chief Pavel Tocci and continued, "When I was informed the ship would be towed in, I asked Pavel to send two of his people to review all the records we could find aboard the ship. According to those records, they had an infestation of something from Eden, but they had no idea what it was or how it got on board."

He paused, then waved a hand towards the life cycle display. "Their records of the infestation are consistent with this…kraken creature. Which supports your theory that the things you found in the brothels came here by way of the *Lucky Strike*! But, just as they couldn't explain how these things got aboard their ship, I can't imagine how the damn things would have made their way from a deep space vessel tethered, out there, to a space buoy, to our red-light district!"

Jane nodded and leaned forward. "I think we can help you there. As I said, we've been reviewing information published by the Eden xenobiology team." Returning to the lifecycle diagram, she pointed at the depiction of the worm stage.

"During the parasitic infestation, the worms pervade the host's body, including their reproductive system. The Eden research team documented an infected male may pass parasites to his sexual partners through their ejaculate. It's not clear whether an infected female can pass the parasites through sexual activity. It's not been documented, but that doesn't mean it's not possible. And, one suspects that parasites could be passed through any exchange of other tissues or bodily fluids."

She turned to Yoshi and continued, "I suspect that you'll find that someone aboard the *Lucky Strike* was infected, either through exposure on the planet surface, or sexually, before the ship left Eden. Either way, I suspect that, once the kraken infested the ship and consumed the crew, they completed their lifecycle and released spores throughout the ship. And, if there were no survivors of the *Lucky Strike* crew to patronize our red-light district, then I can only assume one of the persons that went aboard to salvage the ship was infected by a kraken spore, then patronized the brothels."

"There were only two members of the salvage crew that went aboard," Pavel said. "A harbor pilot..." he paused as he checked his wrist computer, "named Robin Borman, the daughter of the salvage crew's leader, Marcus Borman. The other is Brad Tam. If there were more than twenty infected sex workers, I'm guessing they were both infected and that women can, indeed, pass along the worms."

Jane raised an eyebrow and tilted her head. "The, uh, brothels in question share a reputation for offering a performance enhancement drug. Illegal, of course, because it's actually quite dangerous, but also, allegedly, quite effective and therefore expensive. While you're quite correct, Pavel, that the two may have spread the infestation, it's also quite possible that *just* the male did. Either way, we need to find them both, and quickly, before anyone else is infected. And, Pavel, I'm sorry, but if two of your people went aboard, we need to check them as well. They might have come into contact with a spore while aboard. And we'll need to make sure nobody else boards that ship until it's been cleared of spores."

Pavel blanched. "Shit...well, I'll get you their names and have them report to sick bay. What can you do for them?"

"We have, through our xenobiology department, access to much of what the Eden colony has published. That includes a nanobot surgical procedure they developed to remove the parasitic worms. The sooner we screen them, the gentler the experience will be."

Yoshi cleared his throat. "After notifying Governor Lubandi, I approved refitting the *Lucky Strike* for its next crossing. Kim, we'll need to rescind that approval and determine whether anyone else has been aboard."

She nodded and tapped at her wrist computer. Yoshi leaned forward and meshed his fingers.

"Alright, well, that certainly helps us understand what must have happened to the *Lucky Strike* and what appears to be happening here at home. Now what?"

Jane clasped her hands together on the table and contemplated her fingers for a moment before speaking.

"We're in trouble. Serious, serious trouble. If you go back and read the reports about the Eden colony kraken incident, some fifteen to twenty of these things from one skeletonized host managed to kill 143 of the original 1,000 colonists. From what we understand about the number of these things that hatch from a host and the number of skeletonized victims we found, I would guess, and this is only a guess, that we have some 300 of these things wandering around Alexandria. Maybe more. Maybe a lot more."

She glanced about the room, giving everyone a moment to process that information.

"Now, that's based on the number of victims we know about. We don't know whether anyone infected anyone else at another brothel or in a, uh, non-professional encounter. While it appears unlikely that the female hosts could have infected any of their other clients, we cannot rule that out."

Pavel raised a finger and said, "Now I know why you asked me to track people calling in sick. A few, of course, but at a normal frequency. We'll get names for you so you can follow up."

Jane nodded, pursed her lips, then leaned forward. "We can, and should, presume we have at least 300 of these things loose. If they're following the same pattern as Eden colony, they're spreading through the colony's infrastructure, air ducts and the like, hunting vermin. For now. As they grow, and they grow extremely quickly, they'll move on to pets, livestock, and then people. We should expect reports of missing pets and livestock within a week, then attacks on colonists."

She glanced around the room and continued, "These things are absolute nightmares. They're horrific, voracious predators. They don't sleep, they're never full, they grow very rapidly, and they're incredibly tough. They're extremely clever, smarter than dogs. They have excellent camouflage, matching color and texture of their surroundings to virtually disappear. They have excellent eyesight, smell, and hearing, as well as pits along their flanks that allow them to feel air pressure fluctuations. Most ominously, they have an extremely well-developed electromagnetic sense. They can 'feel' the electromagnetic fields generated by potential prey. They can use this sense to interfere with prospective prey's mental processes; they can put you to sleep, make you feel dopey, even immobilize you."

She rapped the table twice with a knuckle, then said, "My team will work with the Deep Space Academy's xenobiology team to put together some background for everyone to review."

Pavel turned to Yoshi. "With all due respect to Governor Lubandi, we need to prioritize our public safety above her desire to limit distribution of this information." Yoshi glanced around the room, then nodded.

Pavel turned back to the table and resumed his train of thought. "So, in addition to tracking and tracing to understand how many others may be infected and how many of these creatures we're dealing with, I'll also work with engineering to see what kind of weaponry we can come up with to help us find and destroy these things. That'll certainly involve informing the general population for their own safety."

Pavel turned to Jane and asked, "Did you notice whether any of the Eden reports reviewed what they tried? What worked and didn't?"

"Yes, the reports are complete. They were on an orbiting platform, so their use of projectile weapons was constrained. Can't have bullets piercing the hull, right? But, they also found the things were resistant to gun fire. Something to do with their skeletal structure; it's hard to get a bullet through to their internal organs. They relied mostly on flamethrowers. They had some success with their own electromagnetic field generating equipment."

Pavel nodded along. "The crew of the *Lucky Strike* reported some success with crawlers they dispatched through the air ducts to hunt down and incinerate the little buggers. Used a sniffer to detect them from a distance, then infrared to find their heat signatures; finally, they captured the pests and incinerated them. But, my review of their records suggests the poor bastards severely underestimated the growth rate of the things. If we build crawlers, and I think we should, then we should design them to deal with larger predators."

Everyone turned to Engineering Chief Colbert, whose fingers were flying across his wrist computer as he listened. Yoshi, smiling, asked, "Jack?"

Colbert froze, looked up, and blushed. "I'm, uh, waking people up. Yeah. Uh, Jane, Pavel? I'm going to need anything you have on what was used at Eden and aboard the *Lucky Strike*. And, about the critters, see what we might come up with that they missed, or decided couldn't be used aboard a ship or orbiting platform."

Jack turned to Kim. "We're gonna need a lot of help, here."

She nodded and said, "You'll have it!"

Yoshi stared off into space. This couldn't really be happening, could it? Alexandria was home to just over 200,000 colonists. He struggled to process how bad this could be. Eden lost 143 from 1,000. Could they really lose 20,000 of their 200,000?

He looked about the room, wondering how many of his staff had really internalized what they were about to face. They were a good team. Most of them had worked together for years. Kim was a recent addition. She'd arrived via PIP about a year and a half ago; she finished her degree while working her way up the colonial operations hierarchy through fabrication. She'd impressed a number of her superiors along the way, including her predecessor. When he abruptly decided it was time to return to the DSS and took a Chief Engineer position, he nominated Kim as his successor. Generally speaking, her performance impressed Yoshi, but it wasn't a great time to have a Chief Engineer still learning her new job. But, there wasn't a thing he could do about that now.

"Kim, I want you to do *whatever* you have to do to determine whether anyone aboard that *Eureka* could possibly be infested with these worms. I want their captain notified of all this, before they accelerate beyond our ability to communicate, in case someone aboard is infected."

"We could send them Eden colony's nanobot surgical procedure for clearing kraken parasites!" Jane commented. "I'll get it to you so you can send that along to their Chief Medical Officer…"

"It's too late," Kim interrupted. "I mean, we can try, but I just checked, and they're already beyond our ability to communicate."

The room was silent, then Yoshi said, softly, "try. It can't hurt to try. But make it happen ASAP, ok?" Kim nodded.

Yoshi placed his hands on the table. "We'll do our best with respect to the *Eureka*, but our primary focus has to remain our own people. We'll need to communicate with them. I don't want a panic, but we'll need to prepare them for what we're going to have to do to protect them." Yoshi paused, then asked Pavel, "do you think we should arm the general population?"

Pavel, aghast, leaned back. "You realize that most of these people have never touched a weapon? I mean, we have some folks that participate in shooting sports. Archery. Firearms. Hell, we have a few people into martial arts and fencing! But that's a long, long way from combat training! It's one thing to aim at a target down the range, it's another to aim at a, what, an alien monster bearing down on you, all while trying not to accidentally hit bystanders!"

He leaned forward. "My people are all trained to safely handle weapons. Yes, the colony's grown to the point that we have crime, sometimes violent crime. My people can handle themselves, to a degree, when confronted with one or a few armed bad guys. But this is going to be something else!" He waved his arms. "*Flamethrowers*, for God's sake!"

Yoshi stood. "I get it. We're not prepared for this. Well, we don't have a choice. This is happening, whether we're prepared or not. In a few very short days, these…these things are going to threaten our people. We need to arm anyone that can safely handle a weapon. If we have colonists that participate in shooting sports, get their names so we can put a weapon in their hands. Are we taking a risk? Is it possible one of our people will panic and injure or kill a bystander? Absolutely. What's our alternative? We just don't have enough people with weapons training, and we don't have time to do anything about that. But I'm certainly open to better ideas!"

Pavel hesitated, then said, "understood. I don't want to arm anyone without weapons experience either. God help us if they start incinerating bystanders! But anyone that's not armed may as well be on a buffet."

The room was silent again. Eventually, Yoshi said, "I'll defer to you. If you have people you trust to judge the aptitude and attitude of novices that sign up, I'll trust your judgment. Theirs, too. Because, at the moment, I can't decide which is worse. Colonists incinerating each other or huddling in their residences waiting for these things to pick them off."

Pavel sighed. "I have a few people I trust to make such a call. We'll take volunteers, arm the ones we can, keep an eye on them. The rest can serve in supporting roles."

Yoshi nodded. He looked around the room, then turned back to Jack and Kim and asked, "What are our options to lock down the domes? We can't know if these things have spread beyond the primary dome, at least not yet, but we should do what we can."

Kim looked at Jack, then said, "We'll work on that. We're really not designed for it, but we'll see what we can come up with."

"Sounds like we have a plan, or as close to a plan as we can have at this point. I'm thinking we should reconvene in a few hours, maybe, oh, late afternoon?"

Yoshi turned to Jane and added, "Please invite one of your new xenobiology friends."

He glanced around the room as everyone nodded, then said, "I'd best notify my superiors, see whether anyone anywhere has had any great ideas since Eden's kraken incident."

DAY 11

FABRICATION CENTER

Rafael rubbed his weary eyes.

Shift change was approaching. Shift change was always a little stressful. The inbound crew worked alongside the people they were relieving for up to an hour to ensure a smooth handoff. Then they'd get to head home.

Not today, though. It had been a long day, and he was looking forward to going home. For a few days, in fact; his week was ending, and he was looking forward to some time with his family. But, now he wasn't going home tonight. He shook his head.

If you want to make God laugh, tell him your plans...

He looked across the fabrication center at the operations people setting up cots. There wasn't any room for them here on the fabrication floor, so they were reorganizing the satellite rooms surrounding the fabrication floor to make space. Even after all these years, the scale of the place amazed him. He could remember, years ago, when all the fabrication equipment fit in a corner. Now, the space was filled, and colonial planning was sorting through where they might build more interim capacity versus waiting for construction of the second settlement's base station.

Interim capacity wouldn't be easy. Fabrication equipment was loud, wide, and tall, logistically challenging, too. Raw materials harvested around the planet and solar system needed to be brought in, and finished products sent out. This facility was built into the foundation of the space elevator base station. The room was a square kilometer. Beneath their feet, the base station's foundation was a 10 meter thick, 4 km wide disk edged with the stem wall that supported the dome's structural arches. At the center of the room, an immense freight elevator linked them to the gondola station of the space elevator. Directly above them, the inter-dome shuttle tunnels linked the primary dome's transportation hub to every other dome, settlement infrastructure, landing pads, storage silos, and, someday, the next settlement to be built to their west along the equator. There was a thin band of space around the edges of the room: kitchenette counters, restrooms, storage rooms for equipment, materials, etc. Operations folk rummaged through the storage rooms, a little too cavalierly for Rafael's tastes, to make room for cots. Well, it would be good to set the cots up in the rooms. A little less noisy than the fabrication floor.

But making room would be tough. Everything they'd been working on was placed on hold, and all of that stuff needed to go somewhere. But not *here*, because they needed the space for cots. So, the operations people, in addition to setting up cots, carted off anything and everything that wasn't required to support their new fabrication efforts.

He reviewed the templates again. They were creating "sniffers," handheld and machine mountable sensor devices tuned to a set of proteins he'd never seen before. And flamethrowers.

Flamethrowers!!! What the fuck?!?

A machine mountable version and portable incinerators consisting of a backpack mounted fuel canister and a handheld wand, synchronized with their fire suppression systems.

Why machine mounted sniffers and flamethrowers???

For the duct crawlers, of course. Robots with a sniffer, an incinerator, a robotic arm, a camera, an infrared sensor, and a motion sensor. He'd never seen anything like this stuff.

"Why?" he had asked.

"Never mind," he had been told. "Just get started and you'll be briefed later."

Well, he'd worked here a very long time, and, as he ascended through the managerial ranks, he'd learned to trust his leadership. So, if they said to build this shit, well, he'd build this shit. He'd pass instructions along to the next shift leader, then he'd hunker down for the night in his cot.

He wished he could notify his family, though. Their communications links were deactivated. They couldn't communicate with family or friends, at all. And, once again, no one explained why.

DAY 11

STRATEGY

After this morning's emergency meeting, as Yoshi watched everyone file out, he decided to conduct his day's business from here. Then he decided to clear his entire schedule. He realized he wasn't going to be able to focus on anything else.

The next thing Yoshi did was notify the DSS of their situation. Then he went through the material Jane provided about the Eden kraken incident. To call the report "sobering" was an understatement. He sighed, then stood and walked back to the western observation windows. He needed to clear his head. His staff would be returning soon, and he needed to be ready to lead. He closed his eyes and pinched the bridge of his nose for a few moments. Then, with an abrupt shake of his head, leaned against the window.

The view within the dome was impressive. He looked upward through one of the ceiling panels to see a gondola ascending the space elevator: the stalk, as it was commonly known. The dome was fully built out now. Any additional capacity would have to wait for the second settlement. He glanced towards the model in the corner, configurable to depict any colonial development stage: past, present, or future. At the moment, it displayed the current state. There was the primary dome, the nearly fully populated residential dome, ground broken for their second residential dome, a series of twenty agricultural domes with space reserved for twenty more, the zoological park and botanical garden dome, industrial domes, infrastructure domes, their fusion plant, landing pads and storage silos, and, beyond all that, the second base station, under construction and almost ready for the first set of space elevator cables.

So much potential!

If Alexandria suffered comparable losses to Eden, the kraken could take 20,000 souls.

20,000 souls!

That was a staggering, horrifying figure. He shook his head again, then returned his attention to the window.

"Whatcha lookin' at?"

Yoshi turned to greet Jane as Jack followed her into the room. With a forced joviality, he responded, "Oh, you know, just admiring our thriving metropolis! As one does!!"

"Oh? Does one?" Jack laughed as he took his seat, then added, "I've always found that phrasing awkward."

"Would you prefer a 'royal we'?"

"Ha! Uh, no, no thank you. I like our current arrangement just fine. Uh, *sir*. Or would you prefer 'your lordship'?"

Yoshi snorted, then looked at Jane as she made her way to the counter to refill her coffee.

"Jane, did you get any rest today?"

Without taking her eyes from her mug as she topped it off, she tilted her head. "No, not really. But I'll be okay."

He wagged a finger in her direction as she turned and chastised her. As Chief Medical Officer, she, of all people, should appreciate her need to get some rest. On the other hand, she had, on more than one occasion, amazed him with her ability to function with minimal sleep. Kim and Pavel entered as Jane turned, smiled as she offered an obscene gesture, then responded.

"Yeah, yeah, nag, nag, nag…seriously, Yoshi, I appreciate the concern and agree. And I've had a few power naps today. Once we get through this planning meeting, I promise, I'll get some proper sleep."

"Should I set up some cots here?" Jack asked.

Yoshi, walking to his seat, scanned the space thoughtfully. "Yeah, I think we're going to need them. I'd also like you to set up some work consoles. I fear that, at some point, we're going to be here around the clock. I presume you noted, per Jane's report on the Eden incident, that Lubandi and her team found such an arrangement useful?"

Jack and Kim both nodded, then Kim volunteered, "Jane, thank you for preparing that primer. Really useful. Terrifying, but useful."

There was an awkward silence. Part of Jane wanted to acknowledge the complement and give credit to those she tasked to prepare it, but that felt awkward, given the circumstances. Instead, she simply nodded, then informed them that Keith Saputra, head of their Deep Space Academy's xenobiology team, would be joining them soon. He'd descended via landing craft and was making his way to them. Last she'd checked, he'd progressed from the landing field to the transportation hub of the base station and was on his way up.

Yoshi nodded, then asked for updates.

"We found Brad Tam, or what's left of him," Jane reported, soberly. "Confirmed that he's our primary vector. Apparently, as soon as they returned from recovering the *Lucky Strike*, he booked a room in the red-light district and had himself a grand time. We found his skeletonized remains in that room." She glanced around the room. "There were six crew members aboard the shuttle tug that towed in the *Lucky Strike*, and we interviewed the others. They noticed Tam scratching what we now believe must have been a spore entry lesion.

Tam and Robin Borman, the harbor pilot, were the ones that went aboard the *Lucky Strike*. We checked their environmental suits and found that they'd picked up quite a few spores. Which, of course, they couldn't know. Our working theory is that, as they removed and hung their environmental suits in the tug's locker, Tam came into contact with one or more of the spores on the surface of the suits. It's interesting that we didn't find that any of the spores managed to get through a suit. I suspect, based on the Eden reports, that they didn't have time. They weren't aboard the *Lucky Strike* very long."

She shook her head. "All of that's academic, really. One way or another, Tam was exposed and, before succumbing, managed to pass the parasites along to quite a few of his, uh, service providers. The other members of the salvage team are clear, and the environmental suits are being scrubbed. And, we've identified the other clients of the sex workers infected by Tam. None of them, as far as we can tell, infected any of their other customers."

She turned to Pavel. "One of your people was also infected from his time aboard the *Lucky Strike,* but is responding well to the Eden nanobots."

Pavel nodded as Jane turned to Jack. "Now that this infected person is under treatment in my hospital, he won't be contributing any more Eden proteins to colonial waste reclamation systems. I don't know how long it will take to flush his previous contributions from the system, but, once we're through that period, you should be able to use monitoring to determine whether we have any other hosts to find."

Jack nodded. "Should be cleared through by tomorrow based on what I've read from the Eden team. We'll be set up to monitor tomorrow midday, and, at that point, I should be able to tell you whether we've found everyone."

Yoshi nodded, then asked, "what else?"

Kim glanced around the room, then spoke. "The fabrication center set aside all other projects to focus on the designs Jack sent over. We're producing kraken sniffers, Eden protein sensors, duct crawlers, incinerators…uh, and flamethrowers. Some are designed to be mounted on the duct crawlers, others are portable units, backpacks and wands."

She nodded to Jack, who continued. "Reports from Eden showed flamethrowers were the most effective. Projectile weapons didn't work as well. We designed the flamethrowers to sync with colonial fire suppression systems. We need those systems to delay their response long enough for the kraken to die, but not enough to let fires get out of hand."

Yoshi asked, "Please explain the sniffers and sensors? You mentioned the sensors a few minutes ago and I'm not quite clear on them."

"The sniffers are straightforward," Kim responded. "The *Lucky Strike* crew came up with that idea. They smell kraken. Remember, their camouflage is quite effective! When the sniffers are ready, we'll have people go through the colony to get an idea how far these things have wandered. The protein sensors are from Eden colony. If anyone's infected with worms, they'll pass Eden proteins in their urine. If we monitor our waste reclamation and detect Eden proteins, we'll know we have more infected hosts. We'll deploy them at waste reclamation system junctions so we can narrow down where the infected are and get them treated. I think that's our highest priority right now, y'know? To figure out just how much trouble we're in?"

"That's good! What else?"

Jack spoke up. "We're working on these electromagnetic gizmos the Eden colony came up with. Kraken, apparently, have a powerful electromagnetic organ. They use it to sense electromagnetic fields generated by their prospective prey. These devices detect their electromagnetic signals, which helps find them. They can also generate electromagnetic interference that drives the kraken crazy, which could be really useful, especially when they're bigger." He glanced around the room, then added, "if you haven't carefully read about their electromagnetic sensory capability, you really should. In addition to helping them find prey, they also learn how to manipulate prey. Bombard you with these signals, make you dopey, even sleepy. It's actually hard to believe, but the Eden reports are clear. And terrifying! Anyway, we're playing with the Eden design because it's pretty energy intensive. We're hoping we can make them a little more energy efficient."

Yoshi nodded, so Jack continued. "Eden also published processes for detecting and synthesizing kraken sex and alarm pheromones. We definitely want to be monitoring for sex pheromones, and we may be able to use synthesized sex pheromones to attract them to traps. Alarm pheromones might be usable to drive them out of an area. And they developed equipment for detecting and disrupting their electromagnetic sensory capabilities. Fortunately, we have considerable fabrication capacity, so we ought to be able to crank this stuff out."

Yoshi raised a finger, glanced around the room, and asked, "Has anyone heard from Eden yet?" The room was quiet as looks and shrugs were exchanged.

Yoshi nodded, then said, "I'll reach out directly to Elke." He turned back to Kim and asked, "pheromones?"

A new voice piped up. "Pheromones are chemical messages passed with your own species. Sex pheromones attract potential partners for mating, and alarm pheromones alert your fellow kraken that something's hurting or killing your kind. We might find it useful to produce some alarm pheromones to drive younger kraken away. Later, sex pheromones could lure them to traps."

Jane stood and walked over to greet a slender young man who had just entered the conference room. She gave him a smile before turning to introduce him. "This is Keith Saputra, from our Deep Space Academy Xenobiology Department!"

She turned to Keith and asked, "Presume you'll need some lab space?"

"Oh, great, yes, thank you. We, uh, as you might imagine, well, this is new for us, too, right?" He surveyed the room. "This planet is the only life bearing body in the system, and 'life' on this planet, as you may know, is confined to complex organic compounds. Valuable ones! But we don't even have prokaryotic cells, uh, the equivalents of bacteria. So, as you might imagine, most of our local deep space academy's xenobiology department is correspondingly limited. I'm a biochemist, myself. Now, we *do* have xenobiology grad students aspiring to migrate to Eden or pursuing zoological/botanical park careers, so we do have some applicable expertise. And, for our biomedical teams, we have breeding colonies of lab animals. Drosophila, crickets, mice, rats, bunnies, beagles. But we really have limited lab facilities for something like this." He blinked and shrugged.

"Right," said Kim. "We're fabricating kraken compatible enclosures for you per Edenic specifications. As for other equipment, we think it makes sense to build fresh equipment here rather than take the time to bring anything down. We *are* prepared to bring down lab animals. What we need from you is to review and affirm or correct our plans so we get you up and running as quickly as possible."

Keith nodded enthusiastically. "Yes, thank you. And were you able to find us lab space here at the base station?"

Kim smiled. "Right down the hall! I'll walk you over when our schedule here permits, and you can see how it's coming along."

"Excellent, thank you. I understand you have some specimens collected from the victim?"

"Yes, they're already in the lab. They're caged separately; Eden reports they're cannibals."

"Excellent," Yoshi said. "Long way to go, but solid progress. Jane, anything else?"

"Not that I can think of."

Turning to Kim and Jack, Yoshi said, "I guess our immediate priority is to produce and deploy these duct crawlers as quickly as possible. Let's start with the primary and residential domes. I realize we don't know yet whether they've spread to the residential dome, but let's not take any chances."

They nodded, then Jack added, "I have teams ready to deploy them. Let me know when they should show up at the fabrication facility to get the first tranche of crawlers."

Kim promised she would and started tapping her wrist computer.

Yoshi turned to Kim and Jack and asked, "What can we do about isolating the domes to limit their spread?"

Jack leaned back and raised his hands briefly. "Not much, I'm afraid. We're really not set up for that." He glanced around the room, then continued. "Remember, the planet's atmosphere is toxic, and, of course, the temperatures out there are deadly hot, but there *is* an atmosphere. So, we're not worried about an explosive decompression or anything like that. As long as we maintain a positive pressure under our domes, if there's a crack or a hole or anything else, we lose a little air, but we don't let anything toxic in. So, we haven't designed the domes to facilitate isolation!"

He brought up a wireframe model. "Each dome manages its own life support, sure, but they also circulate their air through our agricultural domes and algal farms to harvest our carbon dioxide production and refresh our oxygen levels." He highlighted ductwork. "There are junctions that could be fitted with screens or some other kind of barrier, but we'd have to make the barriers and then deploy them."

He looked up from the model. "My team and I walked through this. We used some of the *Lucky Strike* logs and Eden reports to get an idea how quickly these things spread, compared to when we believe they started roaming, and, well, we think it's too late. We think they've definitely spread to the residential dome and some of the aggie domes. I think we should prioritize the sniffers to assess the situation, then the duct crawlers, then weaponry."

Kim tilted her head and swore softly. "It's actually worse than you realize. It just hit me! For a deep space vessel like the *Lucky Strike*, or an orbiting platform like the Eden colony at the time, all infrastructure is routed through their ductwork. Sure, the things eventually made their way to the core spaces, but initially? They went through the ducts, so duct crawlers made sense. But here in the settlement? Sure, we have ductwork for ventilation, but our infrastructure is plumbed separately. Behind walls. Above ceilings. And some of the buildings are older. Hell, this base station is, what, more than thirty years old? Plumbing bulkheads aren't as clean as they used to be. Why would they be? Who'd care? At least, not until now. I'll bet as many of the damn things spread through the plumbing and electrical conduits as the air ducts." She leaned back and concluded, "there's no point even thinking about trying to isolate domes with screens. There are just too many pathways between the domes."

Jack nodded, his eyes widening as he processed her words. "Yeah...yeah, she's right! Dammit!" He tapped his wrist computer to add the plumbing and electrical conduits to the wireframe. "Infrastructure linkages, and then the transportation tunnels for people and freight, plus their infrastructure. Yeah, I'm sorry, I think we go ahead and deploy the crawlers, they'll take out some of the kraken in the ducts, and every kraken they kill helps. But, now that you mention it? I think we need to shift to making weapons."

The room was quiet.

Yoshi nodded slowly. "Understood. We'll prioritize the sniffers, sensors, and duct crawlers until, oh, end of day today, then start making weapons. But I also want isolation options. I don't mean 'invest time or materials in it', at least not yet, but put a few people in a room and task them with coming up with options. Just in case."

He turned back to Jack and asked, "What else can you do? Can you build robots that can patrol open spaces under the domes? Residential hallways?"

Jack took a deep breath in and rubbed his hands, a little nervous at the massive task ahead. "Let's get the duct crawlers going, and I'll go huddle with some of my people and see what we can come up with!"

The conversation petered out as Yoshi stared into space. After an awkward silence, he turned to Pavel. "You've been awfully quiet."

Pavel did not return Yoshi's gaze. He stared downward, eyes glazed with dread and fear. His hands gripped together so tightly his knuckles were starting to turn white. Finally, he splayed his hands and said, "It's not enough."

Pavel stood and started walking slowly around the table. "I went through Jane's primer, and I've reviewed everything I could find from the Eden incident. I spent some time running simulations with my team. The steps that we're taking? They're not enough."

He stopped, leaned forward, one hand on the table and the other pointing at the wireframe.

"I agree we must assume the things have spread throughout the colony by now. Certainly, the primary and residential domes. Probably elsewhere as well. And they're growing rapidly. In the next few days, we'll start getting reports of kraken taking pets and livestock. A day or two after that? Colonists. And they'll just *keep...growing*!"

He glanced around the room, his words quickening with panic. "If you haven't read survivor accounts from Eden, you need to. These things are beyond horrific. They have the capacity to put their victims into a stupor or a trance before attacking. As they grow, their capabilities expand, enabling them to stupefy more than one potential victim at a time. They like to play with their food. Once victims are incapacitated, instead of stupefying them, they use their electromagnetic abilities to augment their victim's pain and fear, then eat them alive, taking care to keep them alive as long as possible."

Yoshi's eyes darted back and forth between Pavel, who was slowly unravelling, and the rest of his officers as they leaned forward. He knew, all too well, Pavel's dour tendencies. But that didn't make him wrong. And, sometimes, it helped him see what others struggled to accept.

"Okay, Pavel, no argument. I'm scared, too. In fact, we're all fucking terrified! And, if anyone in this room isn't, then you're not paying attention. But what do you *suggest we do, in light of all of this*?"

Pavel fell into his seat, exasperated. "Our natural inclination will be to think in terms of protecting colonists. We need, instead, to focus on killing kraken. I know...I know, these goals should be aligned. And, of course, they are, to a degree. But our simulations and our review of Eden's reports suggest that responding to threats against colonists will distract our people from methodically finding and killing every kraken we can. That means ignoring calls for help. Because, in the end? In the end, if we interrupt our methodical hunting of the bastards to try to rescue people, we'll actually lose *more* colonists."

He looked around the room, then, gripping the back of his chair, started rocking back and forth. "We must hunt down and kill every one of these things. We have to eliminate them, before they breed, before they produce this dragon generation, or, God help us, release spores!"

The room was silent for a few moments, except for the rocking of Pavel's chair. Then Yoshi spoke, quietly but fervently. "I can't argue with what you're saying, but I need a little help translating that into a 'stop, start, continue'. What are you suggesting we 'stop' doing, and what are you recommending we 'start' doing?"

"Forget the Eden protein sensors for now. Odds are nobody else is infected. If they are? Deal with them after we deal with the kraken. Stop building duct crawlers. Deploy the ones we've built! Every kraken they destroy is one less to deal with! But stop making them." He held up his hand and raised fingers. "Number one? Build the sniffers so we can find them. Number two? Build flame throwers. Not projectile weapons. The things are too tough. Flame throwers and get one into the hands of every person that can handle a weapon. Number three? Identify and arm every individual that can safely handle a weapon. Number four? Once we've identified and armed everyone, organize them into squads. These things? If you try to go up against one alone, it'll stupefy you and eat you! We need our people working in squads. Number five? Make some of these electromagnetic scramblers referenced in the Eden reports. And, finally, last but not least, I agree that we may need an isolation option. Get a few people coming up with ideas."

Pavel paused, then looked at Keith. "Did you say you have grad students preparing for zoological park positions?"

Keith responded, slowly, "…yes?"

"Do they have experience with weaponry? Shooting park animals with tranquilizers? That sort of thing?"

Keith could see where this was going. "Yes, yes, we do. And there's staff from the park. Nobody's there now, of course. They're all in their residential dome quarters. I'm getting pings; they're worrying about the zoo's specimens, but I don't think we want them heading over there…"

"If you have anyone comfortable with handling any kind of weapon, please get them over here. We need them on our squads!" Pavel interrupted.

Keith nodded and walked away to follow through.

Pavel lowered his hand. “We should start with squads of five. Smaller squads mean more squads, so we can move through the colony faster. But, from what I’ve read, as the vile things grow, they get better and better at managing suppression of more than one victim at a time. So, I’m thinking that, at some point, we’ll combine squads into groups of ten. But to start? Let’s do five.”

Yoshi looked around the room as everyone internalized what they just heard. Nodding, he leaned forward and said, “Pavel, your reasoning is impeccable. Unpleasant, but impeccable. I’ll plan to address the colony in, what, two hours? Pavel, Kim, and Jack, I want a plan. You’ll start getting volunteers right after I address the colony, so you’d better be ready to organize them. Organize them and arm them.”

DAY 11

ANNOUNCEMENT . .

Rafael was dozing when the door swung open.

"Hey, everybody, sorry to wake you. Governor's making an announcement!"

Rafael swung his feet over the side of his cot and mumbled as he stretched. "Great, maybe we'll find out what the hell is going on!"

He heard a few affirming grunts as he joined the line padding out towards the nearest kitchenette. He looked around the space and saw everyone in the fabrication center was coalescing around kitchenette counters. A meter tall projection of Governor Gulati appeared on each counter. Rafael presumed the same projection was appearing all over the colony.

"Alexandrians, I apologize for interrupting your evenings. But, I need to share some news with you, news that involves the safety of every citizen.

Some of you may have heard about the kraken incident at Eden colony some fifteen years ago, when 143 colonists were lost to Eden predators. Such losses were horrific, but would have been much, much worse were it not for the bravery, ingenuity, and teamwork of the Eden colonists and colonial leadership.

Recently, the *Lucky Strike*, the deep space vessel that brought the initial Eden colonists to their new home returned to our space port. As she approached, we picked up a distress signal, then learned that all hands were lost."

Governor Gulati paused in acknowledgment of these losses.

"The salvage team that intercepted the *Lucky Strike* and returned her to our spaceport discovered that kraken, the same Eden predators that took so many lives all those years ago, somehow made their way aboard the *Lucky Strike*. We've since learned that these predators start their lives as spores, spores that infest their hosts with parasites that emerge as horrific predators. And now, we've recently learned that one of the salvage team members came into contact with one of these spores. He did not know, he could not know, that he'd been infested with these parasites. Now, after taking his life, these parasites have emerged and infested our colony.

And that, my fellow Alexandrians, is why I am interrupting your evening. Even as we speak, these monsters are spreading through our colony. They are small for now, but they are growing. For now, they are hunting the vermin we strive so diligently to contain. But, as I said, they are growing, and, as they grow, they will no longer be satisfied with vermin.

We must find and destroy every one of these creatures. Destroy them before they grow enough to threaten our pets, our livestock, and ourselves. We are asking two things.

If you have experience with any kind of weaponry, any level of expertise, we ask that you use the link I will share at the end of this message and volunteer to partner with our security team, to find and eliminate this threat as quickly as we can. If not, we ask that you stay home, that you lock your doors and stay home until this threat is eliminated.

We also ask for your understanding as we deal with this threat. After this announcement concludes, you'll be provided with links to additional information. Details about how to partner with our security team, details about what happened to Eden and the *Lucky Strike*, details about the kraken, emergency contact information for reporting any sightings of these predators, and, finally, details with respect to staying informed as we go forward together.

In conclusion, I want to emphasize two critically important points.

First, I cannot overstate the seriousness of this situation. I urge each and every one of you to review the information provided at the end of this announcement. These creatures, though now small, are horrific, voracious predators, and they are growing. *Rapidly.*

Second, as serious as this threat is, if we work together, we can get through this. I am confident that, working together, we can minimize losses as we confront and eliminate this threat.

Thank you for your time and attention!"

DAY 12

CRAWLERS

"Show me where the duct crawlers are, please."

Yoshi, arms folded, stood beside Jack as he tapped his wrist computer. Everyone else was busy at one of the workstations Jack's people set up for them. Jack brought up a wireframe of the primary and residential domes, including the infrastructure between them. The rest of the infrastructure spread out but was cut off at the edges of the conference room table.

It was a detailed wireframe, including all the structures under each dome. The ductwork was highlighted in yellow, and the duct crawlers were red. Tiny red dots scurried along a mesh of yellow lines, mostly under the residential dome. At this scale, their movement was imperceptible. Yoshi suddenly had a new and depressing appreciation for just how large the settlement was.

"How many crawlers are currently deployed?"

"Uh, let me check…thirty-seven so far. They made fifty-three before committing our fabrication capacity to incinerators, so we have sixteen more to deploy. We're getting them out there as quickly as we can."

"Add the other pathways. Electrical, plumbing, all that stuff."

Blue lines appeared throughout, and between, the domes. Jack pointed and said, "note that everything radiates from the transportation hubs, here in the base station and under the residential dome, from the circular residential building, under all the recreational spaces, to the central tower. Similar for all the residences under the agricultural domes, by the way."

Residential structures were built in arced modules, eventually forming a massive circle surrounding a common area of parks and other commercial and recreational resources. At the center was the dome's primary tower that anchored the dome's structural arches. It also housed the dome's transportation hub, professional offices, and more commercial spaces. Agricultural domes were divided into wedges, with individual farming residences encircling their central tower.

"Where have we picked up kraken scent?"

Kim's people patrolled the dome with their sniffers set to short range. As they progressed, whenever a sniffer picked up the scent of a kraken, every point within range was highlighted in this display. Jack tapped again. A number of purple clouds appeared in and around the curved residential structure, the central tower, and some of the structures in the open space.

"Shit."

"Yeah. They're everywhere. Good thing we didn't waste time trying to put up barriers. They've already spread throughout the colony."

"Are we killing any of them?"

Jack stood and stretched, then pointed at the wireframe. "Good news and bad news. The good news is that we've killed a lot of them, fifty-one so far. The bad news? If you model the total kraken population using the kill rate, the number of crawlers, how fast they're moving, and the total volume of the domes within which the sniffers are picking up their scent?" He turned to look Yoshi in the eye. "We estimate the total remaining kraken population to be in the neighborhood of 250 to 350. And that doesn't take into account places we weren't able to sniff, like infrastructure between domes."

"Holy shit…" Yoshi suddenly felt nauseous and weak in the knees. But he kept his composure, despite his rising dread.

"Yeah. And they're in the 4-5 kilo range now. About the size of a small dog. We expected this given their growth rate and how long we think they've been hunting vermin."

He shook his head. "Did you go through that primer? They're insatiable and have a phenomenal growth rate. We modeled that, too. Jane was right. The vermin in the duct work won't satisfy them much longer. We should expect sightings tomorrow or the next day. Missing pets, livestock. Kids will be at risk tomorrow or the next day, then all of us."

Yoshi grimaced. "How are we doing with weaponry?"

"We're on track. We're cranking out flamethrowers. We should have one for everyone that's volunteered by tomorrow morning." He paused, then added, "Yoshi, I'm expecting we'll keep churning them out. Spares, replacements, like that. But, I'd like to reclaim some capacity at that point, for building these electromagnetic generators."

Yoshi pulled out a chair and sat beside Jack. "Help me understand what they do and why you want them."

"The Eden colony reports that the things rely heavily on their electromagnetic senses. That's how they detect their prey, distinguish between different kinds of prey, etc. They learn to use these senses to suppress their prey.

Anyway, they built these devices that recognize kraken electromagnetic signals, and that would be better, more precise than the sniffers. And, they said the devices can be used to drive away the kraken, to irritate them and drive them away. That could be really useful in a sweep, to drive them away and towards a squad armed with flamethrowers."

Yoshi started nodding. "Yes, I agree. Please do that."

DAY 12

MESSAGE RECEIVED

Governor Elke Lubandi walked wearily back to the residence she shared with Chief Medical Officer John Schmitt. She didn't expect to see him any time soon. He and his staff would have their hands full for quite a while. She let herself in and walked to the minibar. She took a generous pour of Irish whiskey to her favorite chair, perfectly positioned by her window, great for appreciating the mountains in the distance. She loved that view. She loved this place.

Dear God, for all its horrors, this is such a beautiful world!

Fifteen years ago, their bud of a colony was almost completely destroyed by the kraken. Even the *Lucky Strike* did not survive the ravages of those *monsters*, according to the message she received a week ago.

A week ago...

It had only been a week since Governor Gulati informed her that all souls aboard the *Lucky Strike* had been lost. She remembered George's review of the desiccated carcass they found aboard the ship. Based on field observations and lab experiments, George theorized the ever adaptable kraken employed a different strategy than the one that threatened their colony. There couldn't have been enough vermin to sustain their growth to a size sufficient to threaten humans. When the non-human prey was exhausted, either a few cannibalized the others or they formed a cooperative pack.

Either way, there weren't that many humans to hunt, which would have left the last kraken starving while its dragon young developed. The lack of nutrients in the parent's bloodstream would have triggered its offspring to cannibalize as well. When the last surviving dragon ate its way out of its parent, it would have taken what sustenance it could from its parent's carcass and released spores before dying. The desiccated carcass found aboard the ship would have been the result of pin prick leaks from the spores.

She sipped her drink and gazed through her window.

George didn't get a chance to test his theories once the pineapple crisis roared through the colony. Raising an eyebrow, she thought he'd probably dig into that again. But not right away!

Elke took another sip of her drink. She held the sample in her mouth to enjoy the complexity, then closed her eyes and swallowed. The warmth felt good! She reopened her eyes and looked at the Garcia Mountains. More clouds rolled past the peaks. Soon, the rains would come, and the desert would flush green again, albeit briefly.

She needed rest. She knew the rest of colonial leadership did, too. There were a few pressing issues that required immediate attention; everything else could wait a day or two. Guido would hold after action reviews.

It would take weeks, perhaps months, to work through recommended actions, so waiting a day or two wouldn't matter. She hadn't realized the degree to which their preparations presumed another relatively large predator like the kraken. It nearly cost them everything. The secmeds have been through hell. She knew these things took their toll. She'd need to confer with Ronnie, gain some insight into how all of this was impacting her. The heroism and commitment of the secmeds could not be overstated.

They needed acknowledgement.

They needed care.

They needed rest.

The colony also needed to step up research. Elke doubted George knew how much heat she took for the funding she allocated for his research; but, unlike any other colony, their very survival depended on understanding Eden's life forms. She enjoyed a level of autonomy regarding funding allocations, as long as she met her mineralogical and organochemical targets. But this near disaster convinced her they needed even more research. They couldn't afford another surprise.

Elke paused mid sip. She needed to talk with George about the dual hub proposal. She was afraid to read her messages. Naturally, she sent updates as she was able, but she hadn't read any responses. Elke could imagine what was waiting for her. If the promoters of the dual hub model were calling for the extermination of kraken before, she could only imagine their reaction to the pineapples. They were probably voting for mass exterminations of Eden predators, if not actually terraforming the planet.

She recalled George's observation of the retreating butterflies. She completely understood George's point that the creatures weren't evil. They were truly amoral, not immoral. Right and wrong, good and evil were irrelevant to them. The butterflies were just wild creatures living out their life cycles, and all the tragedies and atrocities of the past week were the result of human greed and her own shortcomings.

Elke didn't share George's devotion to his faith, but she shared his passion for good stewardship. For her, it was a secular thing. What right did humanity have to nearly destroy Earth's biosphere, let alone Eden's? She didn't have George's grounding in ecology, but she understood that Eden's biosphere was a complex web. Attempting to selectively exterminate species like the kraken or pineapples would reverberate through that web, just as attempts to selectively eliminate Earth's wolves, sharks, and other top predators did.

And the prospect of attempting to terraform Eden, especially with only a sliver of understanding what Eden's biosphere offered, wasn't just economically foolish. It was *wrong*.

She remembered talking to John about all this just a few days ago. She told him she couldn't bring George into the discussion. Well, just imagining what was waiting in her inbox was enough to change her mind. She needed an ally, and George's research was popular. He complemented his research with a series of publications tailored for popular consumption and had a devoted following across known space. In a day or so, Elke would sit him down and explain what was going on. They needed to start making plans. And if the DSS leadership didn't like it, well…

She went to take another sip, but realized she'd already finished her drink. After a long pause, Elke poured herself a little more. She stared down at the swirling, deep brown liquid.

Elke confided in John that she was considering resigning. His attempts to reassure her that she wasn't at fault were hard to reconcile with her sense of responsibility. But he also reminded her that Elke also had an important role to play in the responsible development of Eden's resources.

Elke took another sip; her eyes glazed over a report. Nikko had provided preliminary estimates of their losses. He needed a few days to finalize, but his estimates were bleak. Last week, the colony's population stood at just over 17,000 souls. Nikko could account for just under 2,000 deaths, with another

500 or so missing. And, the odds that any of the missing would be found were poor, very poor.

The people would need some sort of ceremony, some way of acknowledging and processing grief.

Oh, damn! That's right! That reminded her that she still needed to organize a memorial for the *Lucky Strike*. And *that* reminded Elke that she'd seen some urgent messages coming in from Alexandria.

Sigh.

Elke didn't want to process any messages from DSS leadership about the dual hub proposal, not yet; but she really couldn't ignore those messages from Alexandria any longer. Elke stood and walked to her communications console. She applied a filter to highlight messages related to Alexandria and the *Lucky Strike*. Probably something to do with processing salvage claims. She knew they could get complicated when cargo and personal effects were involved.

Elke paused, trying to remember if she'd ever heard about salvage claims involving a deep space vessel. After a moment, she shrugged. She couldn't recall any precedents, but this couldn't be the first time it's happened. She'd have to research that. Elke could understand why they'd want *her* involved in claims against the cargo that could be her colony's property. As for the personal effects, well, she'd be honored to have that custom coffee mug monstrosity Moira made for Sri...

Elke returned her attention to the console and started scrolling through the messages, initially irritated that they'd started marking them "urgent". Didn't seem appropriate, given everything she'd been dealing with these last few days.

Not that they know anything about that, of course...

She sighed, began reading, and dropped her drink.

▪ ▪ ▪

Elke sat back, closed her eyes, collected herself, then fetched a towel to address the spill. She returned to her console, emotions cascading through her as she processed the messages from Alexandria.

First, she felt horror, sheer horror as she remembered the "Kraken Crisis" that befell Eden.

Second, she felt refreshed guilt and remorse for the friends and colleagues lost aboard the *Lucky Strike*. It wasn't that long ago she'd learned of their fate. If only they'd been able to warn them!

Third, and swiftly brushed aside, Elke felt a wave of self-pity. Of course, they'd do everything possible to help Alexandria, but did it have to happen *now*? Now, while they were so exhausted from this past week of hell? Thousands lost to another Eden predator, one they should have been better prepared for. It just wasn't…fair? She shook her head. Since when did "fairness" apply?

Fourth, fear. Fear of the implications. She was already deflecting suggestions that they eradicate kraken and related species to allow humans to safely walk the surface of Eden without environmental suits. Such suggestions came from extremist minorities at first, then from more and more as their ordeal with the pineapples progressed. Now Elke could only imagine how the pressure would build with every Alexandrian taken by kraken.

Fifth, and finally, Elke was resolved. The people of Alexandria needed their help, and, by God, they were going to get it. But also, she felt an acceptance that it was time to explain the "dual hub" proposal to George. She was going to need his help assisting Alexandria and she was going to need his help fighting for the preservation of Eden.

Elke sighed. George was still recovering. She should have gone to him, but there wasn't time. She sent John a quick query asking after George's health, then sent word that he should report to her conference room as she prepared to respond to Alexandria.

DAY 12

EDEN RESPONDS ...

Yoshi heard a ping from his console. He walked to his seat at the conference room table and tapped the message as everyone else returned to the table. A projection of a haggard Elke Lubandi appeared above the table.

"Governor Gulati, I just read your latest messages. You have my profound apologies for our silence. We've been through hell."

She clearly struggled to maintain her composure as Elke described, at a very high level, what Eden colony experienced over the past week. They'd learned from their kraken incident. But, when greedy fools smuggled in seemingly innocuous life forms thinking they were decorative novelties, Elke and her budding colony faced a new kind of horror. Their plans and preparations helped, yes, but they were also woefully inadequate. As she spoke, Yoshi and the others sank into their chairs.

Finally, Elke paused, collected herself, and rolled her shoulders back. "We are here, and we will, of course, help in any way we can. I suggest we directly connect your staff with their counterparts here. My chief xenobiologist George will speak with you again shortly. He's…recovering…"

She paused again, then continued. "As you might imagine, kraken have been a high priority research subject, and we've learned a thing or two these last fifteen years. Our lead xenobiologist, George Seid, will share everything we've learned, and we'll do whatever we can to help."

■ ■ ■

George entered the conference room shortly after Elke hit "send."

"Governor?"

Clearly, he was still recovering from his heart attack, suffered during the attack last week. He was aging, of course, as were they all, but he seemed years older than he had just days ago. She did a little mental arithmetic; George should be 116 now. Not that old, really, in fact a bit younger than she was. Elke began to wonder how she looked…

She felt a pang of guilt. She'd spent most of her adult life in command positions: ship's captain, colonial governor, etc. Crises happened, and, when they did, decisions had to be made. Decisions that all too often involved hard choices, weighing lives. She'd demanded too much of him, nearly costing him his life; but, in so doing, he'd saved so many other lives. But she was responsible for his life as well, and, over the years, as they explored Eden together, she'd grown quite fond of him. It *was* good to see him on his feet!

She let herself smile. "George, please take a seat. How are you feeling?"

She was pleased to see him walk confidently, albeit slowly, to one of the conference table seats.

"I'm well, Elke, thank you. Better and better, in fact. John tells me I should be able to get back in the water in a week or two."

"That's excellent news, George. I know how much that means to you!"

George's passions were ecological successions, exploring Eden's oceans, and, of course, preserving Eden, which all too often led to conflict with those tasked to develop Eden's natural resources. Elke had not been familiar with "succession" before getting to know George.

Eden was a dynamic world. The tidal forces of her four moons resulted in disturbances like quakes, vulcanism, tsunamis, and wildfires. Mature ecosystems wiped clean by these disturbances recovered through a succession of stages, inhabited by life forms uniquely adapted to these stages through their life cycles to drive the succession. Kraken were a prime example. Their spores infected the survivors of these disturbances and the scavengers that capitalized on them; then, they voraciously hunted successive stages until producing the next round of spores.

Since reaching Eden, George and his team researched every disturbance they could, including those resulting from the recent quadrilunar alignment that also triggered their recent ordeal. When he wasn't at a disturbance, George was aboard the *Calypso*, their recently launched research vessel. Or, perhaps more accurately, he was beneath the ship, exploring Eden's marine environments.

George smiled, then nodded. "Elke, I can only imagine how exhausted you must be. Please tell me what's sufficiently urgent that you've asked me here."

She smiled again. "George, you're right. There are two extremely important things that we must discuss."

She weighed, for a moment, what to explain first. "About a week ago… about a week ago I received word from Alexandria that the *Lucky Strike* arrived. You may recall, when the kraken crisis started, that our first casualty, Esteban Kay, possibly exposed one of the crew members, Robert Esposito, to kraken parasites. Well, he did. All hands were lost, and the *Lucky Strike* arrived on autopilot."

She could see the news hit him hard, so she pressed on. "I asked them to limit distribution of this information until we could inform our people and plan a memorial. Then the pineapple situation spiraled out of control…"

George nodded. To say the past week had been intense would be beyond understatement. Eden's four moons had aligned. As they did so, seemingly harmless native plants, nicknamed "pineapples" for their superficial resemblance to Earth's pineapples smuggled into the colony as novelties, hatched and released horrific predators.

Elke leaned forward, hands clasped, and continued. "I've just learned today that the salvage team dispatched to the *Lucky Strike* was exposed to kraken spores aboard the ship. George…" she shook her head, then said, "George, they are facing *hundreds* of kraken."

She paused; she could see George was shaken.

"Dear God, Elke, that's…" He shook his head, then leaned forward. "What can we do?"

She nodded again and placed a hand on his. "I need you to contact their lead xenobiologist: Keith Saputra. I understand he's a biochemist; but, they have a zoological park, and his staff includes a variety of specialties. They've reviewed everything your team has published about kraken, except for research that's currently underway, of course. Get your best kraken experts together, connect with him, and do anything and *everything* you can to help. They don't have much time, George. Based on what you've published, their xenobiologists believe the kraken are probably already 5 kilos or so."

George shook his head slowly, then declared. "Yes, yes, of course! Yes! We'll reach out." He hesitated, processing this news, then asked, "Elke, you said two things. What else?"

Elke leaned back in her chair, then stood to refill her glass. She returned to her seat and took a sip.

"George, after we talk, I'm going to meet with the rest of my senior staff. John needs to talk with their senior medical people, Nikko with their operations people, Dieter with their engineers…" She met his eyes. "But, George, first we need to have another conversation, about Eden. I am *not* supposed to share what I'm about to share with you. I'm counting on your discretion. Do you understand?"

"Well, no, I can't say that I do at this point. What can't you share with me?"

Elke glanced toward the window, then began. "There's a proposal under discussion among the most senior leadership of the DSS. They're weighing growth strategies."

Humanity's first colonial expansion was tainted by an unfortunate tendency towards tribalism. After a few generations, lunar, Martian, and belt colonists developed a sense of identity; chafed by remote rule, they asserted and vigorously demanded their independence. The fighting ended but remained fresh enough in everyone's memory at the inception of the DSS to inspire PIP Given interstellar differences, planners feared colonists would not only develop a sense of identity, but that sense would be compounded by linguistic and genetic drift. PIP monitored colonial gene pools and encouraged migration through varying incentives including free education. The result was centuries of interstellar peace bounded by practical limitations set by human lifespans.

Elke sighed heavily and met George's gaze. "As you know, we've limited human colonization to within thirty light years of Earth. There's some concern about what will happen as our colonial populations grow, if we're not able to extend more than thirty light years. We have plenty of room, of course, for the next hundred years or so, but, without a breakthrough like a practical faster than light drive or life spans in excess of our current 150 years, things are going to, eventually, get prohibitively crowded.

So, there's a growing interest in what's being called the 'dual hub' model. If we were to change our plans for Eden, if we were to take the steps necessary to allow humans to walk the surface without environmental suits by exterminating kraken and their relatives, if we were to expand our current population targets significantly, Eden could rival Earth. We would then have a 'dual hub', with panmictic incentives churning colonial populations within thirty lightyears of either planet, greatly expanding habitable space."

George blanched. He leapt forward. "Elke, no. No…NO! No, we cannot allow this! This would devastate the Eden biosphere! This cannot be!"

Elke extended her hands in sympathy. "I agree, George. Fully! And I've been fighting it. But, the Alexandria situation just made our lives much, much harder. Consider what would happen if a significant number of Alexandrian colonists were lost to kraken! Public support for the eradication of kraken would skyrocket!" She raised a hand to silence his objection. "I know, George, I know. Eradicating kraken and related species would have profound consequences. But most decision makers don't understand that, and I am very concerned how this will play out in popular discourse."

Shaken, George said quietly, "What do we do?"

"The first thing we do is help. You get your team together, and you work with this Keith Saputra. Do everything you can, without risking your own health, mind you! Help them minimize their losses. I'll work with my staff to send them everything we can. Engineering, operational, medical stuff, whatever we can. That way, you and I can partner to also address this dual hub proposition together."

Elke paused, then added. "Then, my friend, we start by educating everyone about Eden. You cannot, and I must emphasize this, *you cannot speak directly to the dual hub proposal.* You're not supposed to know about it, and I could get into a great deal of trouble for sharing it with you. But, as you decide what to publish for popular consumption, you need to keep it in mind. We need to rally popular support for the preservation of Eden's biosphere!"

▪ ▪ ▪

George was surprised to find himself outside his quarters. He still maintained personal quarters in the surface settlement's base station, a perk from his years as the colony's lead xenobiologist. But, until this crisis, he hadn't spent much time here. He spent as much of his time as possible aboard the *Calypso*.

He shook his head and sighed. Since leaving Elke, he'd been in a fog. How much of that was lingering recovery from his medical misadventures versus the shocks of what he learned from Elke was anyone's guess. And, ultimately, it wasn't that important.

George had left her conference room and made his way through the base station, his mind wandering, paying no attention to where he was going. He wasn't sure how or when he reached his destination. He cleared the portal, entered, and took a seat by a window facing the Garcia Mountains. He needed to process everything he'd just learned. He felt he was being tested.

Hell, they were *all* being tested!

George had never been to Earth, but he was pretty sure that, if he made his way there, it would break his heart. God created such beauty. Grand things, to be sure. The mountains to the east, the ocean beyond, the sky and heavens above them, all were beautiful. But there was *so much* beauty all around them, so much wonder, all the time, if only one took the time to look. The smallest leaf, the tiniest insect, such marvels to behold!

He could see, in the distance, rain falling on the mountains' rain forests. A perfect example of nature's wonders. What could be more mundane than rain? A simple, straightforward, thoroughly understood meteorological process, yet magical to witness.

What a wonder that water *can fall from the* sky*!*

George sighed dreamily. He loved it here. It was a dream come true. No, it was more than that. Eden exceeded all his dreams. It was a challenge, sure. It was a true wilderness, and true wilderness came with tests that were, at best, indifferent and, at worst, hostile to her visitors. But such an opportunity! To be the first lead xenobiologist for the first complex ecology beyond Earth. He leapt at the opportunity and accepted that he'd done so for selfish reasons.

But, once here, he began to understand and appreciate God's plan.

George was a man of faith. He'd never been particularly preachy; he never took Scripture literally. He envisioned God as a mostly "hands off" deity. Once things were set in motion, as those blessed with free will made their choices, God rarely intervened. More than anything else, his faith fed his belief in stewardship. Having blessed humanity with Earth, having witnessed our mismanagement of our home world, that God would grace humanity with a second chance filled his heart and soul with wonder, gratitude, and humility. He believed God wanted him, and wanted humanity, to appreciate, and preserve, the wonders of his creations. And George thought he'd been ready.

Sure, he was familiar with how the DSS developed natural resources. Such development pulled humanity into space as certainly as the devastation of Earth's ecology pushed them out. Intellectually, George understood. Humanity needed resources, and the same tidal forces that drove the dynamism of Eden also drove formation of ore deposits. As lead xenobiologist, it was *his* role, *his* responsibility to evaluate development proposals, to protect Eden to the best of his ability.

Emotionally? It was devastating.

He'd not been prepared for the aggressiveness of the development proposals. Had humanity learned nothing? He'd recently seen some materials distributed by the Edenists, a new ultraconservative religious movement opposed to any Eden development. It struck a chord.

They suggested the kraken were a test, from God, for humanity. A test, a warning, perhaps even a punishment. Punishment for our hubris, for crossing trillions of kilometers of interstellar space in order to impose humanity's selfishness, greed, and myopia on yet another world. The Edenists thought it a test they were failing, and that the kraken were a punishment intended to force the colony to return home, where all humans belonged. Lord only knows what they thought after the pineapples!

As a scientist, George thought this was all nonsense. The kraken were predators, no more, no less. It was no different than our ancestors who, long ago, crossed oceans and continents and confronted lions and tigers and bears. Predators exquisitely adapted to their environment, an environment that primed them to munch on his fellow colonists. Tragedy, sure, but not judgment.

But, George increasingly felt like they might have a point. And, if they're right, then this dual hub proposal would be more than greedy and short-sided. It would be evil.

DAY 13

PREPARATIONS

Rafael supervised loading another suspensor pallet with flamethrowers. Overall, it was going as well as could be expected. There'd been a recruitment call for anyone with any level of experience handling weapons, and it turned out that the fabrication center had a shooting team. Who knew?

Enough of them left that they had to restructure from three overlapping nine hour shifts to two overlapping thirteen hour shifts. Rafael sighed. Tired people made mistakes, so tired people building flamethrowers could get interesting!

Couldn't be helped, but it was going well. They'd finished the duct crawlers and the sniffers. They should be finished with flamethrowers sometime tomorrow, then they'd start making these electromagnetic gizmos. Rafael had no idea what that was about. He didn't care much. All he could think about besides his current task was his family, and whether they were ok.

▪ ▪ ▪

Pavel looked around the auditorium. He'd been here many times as a patron, attending a concert, or watching a play. But he'd never stood on the stage before, nor had he ever seen a pallet of flamethrowers on a stage.

Somehow, it didn't feel real. He knew it was, and yet it still felt like he'd wake up soon, bemused by this dream. He glanced at the time, then at the people still filing in. He told his staff that he wanted to start at 19:00; but, realistically, it was looking more like 19:15. That didn't matter. What mattered was forming squads, and they didn't have much time.

Pavel walked to the edge of the stage and crouched, looking down at Bogdon, one of his security leads. "I'm thinking we give them another five minutes or so, then start. We can do this in shifts."

Bogdon shrugged. "Sure, we'll close the doors, and I'll wait in the hall as stragglers accumulate. When you get the first wave to the breakout rooms, we'll bring in the next wave."

Pavel nodded, then returned to center stage as Bogdon walked to the back of the theater to brief his peers. He turned and gave Pavel a "thumbs up," then stepped outside as doors closed.

Pavel cleared his throat, pressed his hands together, then addressed the room.

"Thank you for volunteering to protect our community!"

▪ ▪ ▪

Amy waited in her breakout room. She could hear Pavel's speech. He was a good public speaker. Went with the job, she supposed. She was sitting on a stool, waiting for the speech to end, after which her first set of heroes would be led into this room and quite a few more just like it. To her right sat a pallet of flamethrowers. Beyond them, a stack of steel boxes stacked on flameproof matting that they could use as a practice target.

She was new to the colony. She'd arrived last year. It felt weird to make the crossing with a bunch of kids leaving home for the first time. She remembered such a time, long ago.

Bunch of green kids!

There were a few older folks like herself. Well, maybe not old, just not kids anymore. They all had stories, stories more interesting than the kids. All of them looking for a fresh start, a second act, as it were. In Amy's case, it was a realization that, now that the kids were grown and gone, she and her partner had drifted apart. So, she signed up for the next deep space PIP crossing, completed a master's degree, and took the first job offered to her.

That "first job" turned out to be the colony's premier shooting range. Amy had never been interested in such things; but, under the circumstances, she gave it a try. To her surprise, she enjoyed it and was rather good at it. Amy was, in short order, competing successfully in marksmanship competitions and serving as their firearm safety instructor.

Which is why she was waiting in this room.

She'd heard the governor's speech. She'd reviewed the supplementary material, was terrified, and volunteered. In a few minutes, she'd be interviewing militia candidates, ascertaining their level of weapons proficiency, and walking them through how to use a flamethrower safely.

A flamethrower!

She shook her head. Things were about to get very interesting around here.

DAY 14

FREEDOM

When it first emerged, the kraken sensed the presence of many siblings. They were not predators nor competition nor prey. They were all about the same size; and, for them, at that size, the host seemed an infinite resource. They shared their host, feasting and growing.

As they grew and the host was consumed, its life signs faltered. The siblings each realized that their hunger would soon require other prey. They each re-evaluated their siblings as potential predators, competitors, and prey. The kraken sensed that it was one of the largest. Briefly, it considered positioning itself to prey on its smaller siblings. But it knew that would be a finite resource. It needed to hunt other prey.

Sensing air movement, it sought the source. Along the way, it encountered smaller siblings and sacrificed them for the sake of its own growth. It came to a mesh separating it from a darkened space. It sensed prey in that darkened space, and it sensed that some of its siblings went through this mesh in pursuit of that prey. But, it had grown too much and could not fit through the mesh.

It weighed its options.

Some of its siblings remained in the vicinity, but they'd also grown. The kraken knew that, as they grew, their capacity to injure each other grew as well. It decided that the potential nutritional value of its remaining siblings did not outweigh the risk of an injury that might make it vulnerable to the next sibling it encountered. Until and unless it came across a sibling small enough to be safely taken, it would, henceforth, avoid its own kind. So, in order to find prey, it had to leave.

But how?

The kraken wandered again, this time more widely. Then, it sensed moisture. It moved towards the moisture and found, in one of the vertical surfaces, a hole. A long, thin pipe carried water through that hole. The hole was irregular and the edges thin and fragile. In the darkness beyond the hole, it sensed prey. It adjusted the edges until it safely fit through into the darkness beyond.

Once through, the kraken explored. Following the piping, it encountered prey. It feasted on roaches; then, as it grew, it targeted rats. Roaming and exploring, feeding whenever it could, the kraken experimented with its prey. It learned how to manipulate them, how to enjoy them, and how to share the exquisite experience of their consumption with them.

As it roamed and fed and grew, the kraken also sensed more and more of its own kind wandering through the same darkness. At worst, predators. At best, prey. But, with so much prey to be had, why risk it? Why not press on, feeding and growing and reveling in the pleasures of feeding on vulnerable, delicious creatures?

The kraken fed and grew, roamed and explored towards prey and away from its own kind, always following the pipes as they merged to form larger pipes. But, over time, as it grew, available prey became less and less satisfying. It would stun such prey, suppress them into stupor, then snatch them up using its tongues. It was sustenance, but not satisfaction. Eventually, after a long, long dark stretch, the piping branched rather than merged. Sometimes, thin pipes went through holes to brightly lit places rich with the scents of all sorts of prey. Some were like the host; but, as much as it had grown, such prey remained tantalizingly out of reach. Then, the kraken sensed other prey, new prey, tantalizing prey, prey larger than the vermin with which it shared the space behind the walls.

So exciting!

When it could, when the prey like the host were not present, the kraken would force its way through the holes that accommodated the pipes. It would cautiously explore, searching for novel prey. It found all sorts of things: delightful, delicious things. Some were bound in cages that it tore apart; the prey delightfully, fruitlessly tried to escape, pressing themselves against the interior of the cage as the kraken leisurely reached out to take them and slowly, carefully, consume them. Others roamed freely. There were two kinds, each varying in size and, to some degree, shape. But both forms, the cats and the dogs, had sharp teeth and claws, though the cat's claws were sometimes deceivingly sheathed. The dogs varied more in size. Sometimes they were very small; sometimes they were challengingly large. But, the dogs seemed unable to climb, while the cats climbed adeptly, making them harder to access.

The kraken learned to suppress the cats and dogs deeply before striking, to ensure its claws gripped them carefully to manage their teeth and claws. But, as soon as they were sufficiently subdued, the kraken everted its digestive organ against them, leaving only their sensory organs exposed. Then it bombarded them with sensory augmentation to maximize their experience. They were so delightfully emotive! So delicious in every way!

The kraken feasted. And it grew, it became bolder and bolder, until it smelled death.

Alarm pheromones wafted through the air! Alarm pheromones signaled something was killing their kind, wafting from the dark spaces behind the mesh panels. Something was hunting them. Something was killing its kind. It could smell their fear, their deaths.

What was it? Was it safe to go through the openings? To take more of the pets? Was there a choice? What was the alternative? Something was killing its kind, and so it had to feed.

It had to grow.

DAY 14

CONSULTATION . . .

"Y'all need to see this."

Keith walked from his team's console to the conference room table. As Yoshi and the rest of his staff followed, he continued. "We've been working with the Eden xenobiology team. We explained what we'd gleaned from their published research, and they've been sharing *so* much more! Ongoing research they haven't published yet, firsthand accounts from their kraken outbreak. It's a *lot*." He glanced around the table. "As you can imagine, it's a hot topic for them. Everything from basic biology to their latest medical procedures for treating parasitic infections to ways to mess with their electromagnetic capabilities. We've been sharing what we're getting with medical and engineering." Jane and Jack nodded their appreciation as he continued, "but, I thought this summary was worth sharing with all of you."

He tapped his wrist computer, and a meter tall projection of George Seid appeared over the conference room table. The projection spoke:

"Governor Gulati, Keith asked me to prepare this overview for you and your staff. Let me start by expressing our empathy for the challenges you are facing, and our commitment to help you in any way we can. To that end, we've provided deep dives for your medical, security, and engineering teams based on our experiences and ongoing research."

Yoshi glanced at Jane, Jack, and Pavel. Pavel held up a thumb.

"The creatures you're facing," the message from George continued, "belong to a diverse and very successful group, adapted to a variety of Eden ecosystems. As I understand, you've already been briefed on their basic biology. The alternation of generations lifecycle consists of spores, worms, haploid kraken, and diploid dragons that produce the next round of spores. If you'll indulge me, I'd like to discuss another level of detail, which I think will help you understand why these kraken are so well suited to plague us, and, I hope, help you prepare to deal with them.

Eden is a dynamic world, regularly disrupted by events like quakes, volcanos, tsunamis, and so forth. When these events wipe out an ecosystem, it recovers through a process we call succession. Imagine, if you will, a coastline has been wiped clean by a tsunami. Everything that *was* living there is gone, and the now barren landscape is littered with carrion. Scavengers arrive. Various things sprout from the Eden equivalents of seeds and eggs and spores, including kraken. Spores are a cheap way to reproduce, and they're everywhere. The kraken that generated the spores cannot 'know' what life forms might be found as the ecosystem recovers, and remember, Eden has five distinct life lineages, each as alien to each other as any are to Earth!

So, the spores parasitize whatever's survived the disruption or arrived afterwards to scavenge; then the spores emerge to hunt as kraken. They're exquisitely adapted to capitalize on whatever's available as the ecosystem recovers. They have amazing digestive plasticity. They can eat *anything*. Kraken are the ultimate pioneers. They're incredibly adaptable. And they grow fast. They have the highest feed conversion rate we've ever seen. They never sleep, and they're never satiated. Their whole existence is a race to grow as rapidly as possible, at first to avoid being cannibalized by their peers, then to reach reproductive size. They're amazingly tough!

We've had success with flamethrowers. In our most recent challenge, we found we could apply small fragmentation charges designed to disperse shrapnel through the predator's skeletal structure, which we think should work with kraken. But, if you don't manage to kill them, they heal very quickly with amazing regenerative capabilities. All of their senses are acute but, as they grow, they become increasingly dependent on their electromagnetic senses, which are, frankly, amazing. And worth dwelling on for a little bit.

They use their electromagnetic capabilities in two ways. First, to detect their prey. Second, to stupefy their prey. Remember they're smart, about as smart as a dog or octopus. They solve problems, and their claws and tongues provide dexterity. They can open boxes, remove panels, things like that. As they hunt, they experiment. Earthly prey is new to them; they cannot apply any instinctive knowledge of Earthly life form anatomy, physiology, anything. And yet, as they hunt, as they feed, they learn. They learn how to use their electromagnetic capabilities to suppress prey's thinking, make them dopey, put them to sleep, and paralyze them. And their ability to manipulate their prey becomes increasingly sophisticated.

They are also unconscionably cruel. They take satisfaction from their ability to keep their prey alive while feeding, and from using their electromagnetic capabilities to augment their prey's fear and pain. We've documented more mature kraken manipulating multiple targets at once, keeping prospective prey immobilized while maximizing the suffering of the prey they are consuming."

George visibly shuddered as he added, "I recommend you review some survivor accounts." After a pause, he continued. "So, the kraken generation is incredibly flexible and adaptable, designed to grow as rapidly as possible. Once they reach reproductive size, they seek mates. They are hermaphrodites; any two kraken can impregnate each other. But remember, they are cannibals, so their courtship includes exchanging prey. If you don't want your lover to eat you, hand them a meal!

Incidentally, and, as a testament to the adaptability of the kraken generation, this apparently did not happen aboard the *Lucky Strike*. Based on what was found on the ship, and consistent with what we've observed in the field and lab, we think when the kraken exhausted smaller, non-human prey, but were not yet large enough to take human prey, they formed a pack. Then, when the last humans were consumed, we think they mated, cannibalized, and the last kraken produced one dragon that immediately dispersed as many spores as it could produce. Which brings us to the dragon generation…

Kraken are, as I've said, highly adaptable. Their mission is to gain mass as rapidly as possible to breed as soon as possible. When they breed, they produce dragon offspring. But, while the kraken is the ultimate generalist, their dragon offspring are exquisitely designed to exploit whatever environmental conditions they'll face. How?

Well, first, the kraken reproductive system is very different from ours. Each individual generates egg cells like ours, but the sperm they exchange do not contain complete genetic packages. They contain fragments, and, as they continue to mate, they acquire a broad portfolio of genetic material. Now, all of your kraken are descendants of the one spore that infected Brad Tam, so they're genetically identical, but the resulting sperm are *not*. Kraken, through mechanisms we do not yet understand, can tailor the genetics of their dragon offspring! Then, everything each kraken parent has experienced, and continues to experience through gestation, influences development of the dragon generation. Stress hormones. Alarm pheromones. Quantity and consistency of blood borne nutrition reflecting their hunting prowess, available prey, and competition. The result? Every kraken that survives to reproduce will produce its own version of the dragon generation, optimized for whatever the parent experienced. And those dragons, from the moment they eat their way out of their dying parent's body, will be just as smart, just as cruel, just as fast growing as their parents. Voracious, insatiable hunters, doing whatever it takes to grow as rapidly as possible to release as many spores as possible."

George paused again, glancing about, almost as if he were making eye contact with Yoshi and his staff.

"We don't understand it all yet. It's been fifteen years, and we've learned a lot. But, there's so much we still just don't understand. We survived them, though it cost us dearly, and we will share everything we can to help you through what's coming."

DAY 14
PETS

Yoshi stood by the window again, his mug steaming, admiring his view of the surface settlement as the circadian cycle shifted from day to night. He loved dawn and dusk. Natural light waxed or waned as the sun (albeit obscured by the maelstrom) hovered at the horizon. Colonists increased or decreased the artificial lighting of their spaces to suit their whims as people have since the discovery of fire.

The dome's lighting was provided by an arc of lights sequenced to match the sun's daily, seasonally adjusted progression. Humans managed well enough on orbiting platforms or under domes on lifeless rocks, but the perpetual storms of primordial worlds were profoundly depressing. Early colonies on such worlds found that farmers, living under agricultural domes with lighting designed for crops, fared better than their urban counterparts. Such lighting was now standard for all domes. This created brief intervals at dawn and dusk, during which, if one cared to look up through the skylight, the space elevator's lights gleamed against the darkness.

Just beautiful.

He sighed. They weren't ready for this. Here, among the colonies at the edge of humanity's deep space exodus, there were no conflicts. Was there danger? Of course. Space was a hostile environment. Accidents happened. Tragedies occurred despite their diligence, but no conflicts.

Tensions lingered among the nations of Earth despite emigration's gradual reduction of the global population. Tensions lingered among the lunar, Martian, and belt communities that once fought fiercely for their independence, but that was centuries ago. Now, they were all thoroughly intertwined culturally and economically. Time and resources were still wasted maintaining offensive and defensive capabilities; but, this declined ever so gradually, year after year.

Not beyond the boundaries of humanity's home system, however. Out here, at long last, humanity moved past organized violence. Centuries of peace after millennia of violence.

No alien adversaries.

No beasties bigger than a bacterium.

Until now...

Would people remember how to fight? Elke's people figured it out. When the time came, they fought. They'd fought, and they'd prevailed. Yoshi sighed. His people would, too, because the alternative was unthinkable.

This all felt so strange. He could not help but wonder whether this was what yesterday's wartime leaders felt. When invasion was imminent, when hostile forces gathered to lay siege. He turned to scan the room. His department heads were doing their jobs. Everything was running smoothly. He wondered whether they were too busy to ponder the things he was pondering. If so, he envied them.

Yoshi looked upward in time to watch the stalk's lights come alive, then heard Pavel utter a short barrage of obscenities. He turned to see Pavel step back from his console and walk hurriedly towards the conference room table.

"Pets! I have confirmed reports that they're eating pets. Dogs, cats, you name it!"

Pavel brought up a wireframe display of the residential dome as the rest gathered around him, then highlighted residences in yellow. There was a clear correlation between affected residences and the purple clouds.

"Any sightings?" Jack asked.

Pavel shook his head. "No, not yet anyway. It's so goddam frustrating. If they haven't volunteered to join a squad, they're supposed to be hunkering down and securing their entries. But, there's still people out and about, running errands or whatever. We don't have personnel to spare to send them home. And when they do go home, Fido's missing. Or they find a bird or reptile cage destroyed. My people are walking them through inspecting their residences and they're finding the wall around pipes and conduits torn open. We knew there were kraken roaming the ductwork because the crawlers have been nailing them, but now we know they're definitely roaming the pipes and conduits as well."

Kim frowned, then snapped her fingers and pointed at the wireframe.

"The older sections!!"

Jack paused, then nodded his agreement as Yoshi asked her to explain.

"The older sections of the residential building. Remember? Once the dome was completed, we started building the residential building. We built it stages, right? We built the original units at full height, then, as the population grew, we added structures to either end of it. Now the arc is almost a circle. And, if you look at where the pet attacks are, they're in the older sections."

"Which makes sense because that's where the material around the plumbing and telecom services would be the weakest," Jack added. "It's easiest for them to break through there. We should definitely have some of our squads patrolling the older sections!"

Pavel looked at them, then nodded. "Will do!"

"Any idea how big they are?" Keith asked.

"No, not really. The wall openings suggest they're about as big as a fair-sized dog, but that's a guess." Jack offered, "we're not seeing anything that large in the ducts!"

Pavel shrugged. "Don't know what to tell you. I don't know why they wouldn't be as big in the ductwork, but I also don't know why they'd tear open bigger holes than they need to. And some of the dogs they're taking were pretty big. So, we think they're in the 10-20kg range."

Keith whistled. "Well, Dr. Seid told us they grow fast! I'm wondering whether they had access to more vermin."

Jack responded. "The pipes carry water, and some vermin chew on telecom insulation, so yeah, they likely had a richer food source. But, at some point, they outgrew the vermin. They just don't offer enough nutritional value to justify bothering to consume them. That's why they're going after pets, and we should expect them to threaten colonists soon!"

Yoshi nodded grimly. "So, what do we do?"

Jack tilted his head. "We can analyze the attacks on pets, look for patterns, try to predict where they'll strike next and dispatch squads to be ready for them."

"You know what else?" Pavel added. "We could modify the security monitoring systems. For the public spaces, at least. They pick up violent or agitated behavior. Pings security and we take a look to see if anyone needs help. We could take a look at that monitoring, see what we could do about picking up kraken attacking people."

Yoshi nodded at them. "Do it!"

DAY 15

PREY

Rachel was anxious, and miffed. Just how the hell did this happen? This just wasn't *normal*!

Alien predators?

She'd heard the Governor's address, and she'd seen the recruitment calls for anyone with weapons experience. She knew, just knew that at any moment they'd lock down everything when these nasty alien critters made their appearance.

She'd never heard of such a thing!

Well, she kinda, sorta remembered hearing about the Eden kraken thing. That was *years* ago! Rachel had been a grad student, a "pipper" just arrived from her home world to pursue a degree. It was just a bunch of news stories that she largely ignored. Sure, it *sounded* awful, but it didn't affect her, or anyone she knew, so…

And now they're here? How the hell did that happen?

She wasn't happy, and none of her friends were happy. Well, she'd better walk the dog while she could. They'd be locking everything down sooner or later. Probably sooner! And, once they did, the damn dog was going to drive her nuts. She lived alone in a small apartment on the ground floor of the residential structure. It was cheap! But the dog was a shepherd mix: sweet, smart, active, and big. Too big really for her space, but she'd fallen for the pup a year ago and convinced herself it wouldn't get *that* big and it wouldn't shed *that* much…

Rachel went to the portal and picked up Bunny's leash, which made the dog go absolutely nuts. She really needed to make the time for more training or puppy obedience school. Not that Bunny was really a puppy anymore. She clipped the leash to Bunny's collar, led her through the portal, and sealed it behind her. At least she didn't need to take the lift. The ground floor had crappy views, but it was really convenient for walking the dog.

The two stepped through the outward facing doors and crossed the running track surrounding the residential building. Once they crossed the track, Bunny lost all discipline, sniffing and snuffling at everything that caught her fancy. Rachel indulged her, meandering with the dog, enjoying the thick undergrowth beneath the stem wall and admiring the atmospheric cauldron on the other side of the dome wall. She was in no hurry and idly pondered which door they'd pick to re-enter the building. The ground floor corridors were wide and pleasant, with benches and small plantings, fountains, and play structures.

Bunny froze.

Rachel nearly bumped into her. Puzzled, and feeling a little disoriented, she cocked her head to look at her dog. She heard a faint rumble from Bunny's chest. Her ears were up, as was her tail. Then, suddenly, her demeanor shifted. Her tail tucked, her ears flattened, and Bunny cringed.

Weird. Must be something scary in the bushes? Are there bees? Sometimes there are bees, but the shrubs don't have flowers, so why would there be bees?

Rachel shook her head. Why was she feeling dopey? Maybe it was time to go home.

"C'mon, Bunny, that's a good girl! Let's go!"

Bunny wasn't the most disciplined pup, but she usually responded when called by name. Not this time.

Weird!

Bunny remained frozen, ignoring Rachel's words, not responding to tugs on her leash. Meanwhile, Rachel felt increasingly odd, tired, fuzzy. She realized her vision was narrowing. She realized Bunny's recalcitrance wasn't a problem, not really. It was a pleasant place to stand. There was no hurry.

Rachel turned to look for a bench to sit on, then realized that she hadn't actually turned. She was still standing there, looking at Bunny, one part of her mind believing she was turning and the rest of her mind realizing she wasn't.

Something emerged from the bushes.

Rachel realized that, before it moved, she hadn't seen the creature. Amazing! It looked like the bushes! But, as it moved towards Bunny, bright colors flashed along its flanks. It was the strangest thing! It was as big as Bunny, with four stout legs, two tails, a pair of claws, and, sprouting from behind its beak, a spindly arm ending in six delicate fingers, each ending in an eye.

It slowly but steadily extended its claws, unfolding them from beneath its body, to grasp Bunny. One claw closed about Bunny's muzzle, the other on her hips. One of the eyes studied Rachel while the other five stared into Bunny's. The beak opened and three tongues extended; it grasped Bunny's hindquarters and pulled her legs into its beak. Slowly and steadily, it pulled Bunny into its beak, tongues exploring her body, the one claw firmly holding her muzzle while the other shifted to help the tongues pull the body into itself.

As Bunny's hindquarters entered its beak, Bunny slowly revived. She began to whine, then squeal. She trembled and squirmed, then howled as she struggled, unable to escape. Bunny's eyes broke away from the kraken to look at Rachel, pleading for intervention, then back again as she steadily disappeared into the kraken's gullet.

Rachel, horrified, could not move. Tears rolled down her cheeks. At some level, she knew Bunny was suffering and terrified, but there was nothing she could do.

The kraken's claw held Bunny's muzzle firmly despite her writhing until the light left Bunny's eyes. The kraken's tongues stretched out, pulling Bunny into its digestive organ more rapidly as its eyes turned towards Rachel. As Bunny disappeared, the thing turned towards her, its eyes staring into her own, the claws reaching out for her.

She felt herself fall slowly backwards as the wet, slimy tongues explored her body. She felt her clothing tear away, felt the smooth muscular tongues obscenely explore every bit of her body. The claws held her torso as the tongues slid across her belly and genitals; then, like a thousand knives, they pressed into the flesh of her legs as they were pulled into its stomach. Her feet were pressed into rhythmically contracting moist muscle, then her ankles, then her calves.

As her mind began to clear, she stared up into the thing's eyes. She was horrified and repelled. Then her feet began to burn. Rachel realized she was being slowly digested alive, and, as she realized what was happening to her, she began to scream. And the waves of bright colors along the kraken's flanks pulsed more vividly and rapidly.

▪ ▪ ▪

Dario didn't know what to make of the announcements. How could such things happen? Nothing he could do about it, of course. But he had faith in the authorities and was confident they would have it all under control. He had to have faith. He didn't have the bandwidth not to! Not since his partner passed away, leaving him to be sole provider for their toddler. She was wonderful, absolutely adorable, but such a handful! Kayleigh was his reason for being, and taking care of her was a huge part of working his way through his grief. He understood, from all the resources he'd consulted, why she did the things all toddlers do, but, in the moment? In the moment, understanding did little to ease frustration!

All, well this too shall pass.

But, under the circumstances, the prospect of a lockdown was just too much to bear. He decided they should have a tiny adventure, a stroll to the ice cream parlor for a treat. Holding Kayleigh's hand as they made their way from the residential building along the boulevard that led to the central tower, it was hard to believe it wasn't just another ordinary day. Well, it wasn't *quite* an ordinary day. There were some people about, but not many. Some were very quiet; others engaged in vigorously animated discussions.

He found himself questioning the wisdom of this outing. But the ice cream parlor wasn't far. It was a small, colorful structure set off the boulevard, surrounded by a variety of shrubs. They turned from the boulevard, walked to the parlor's entrance, and found it empty. Open, but empty!

It didn't seem like Pietro to forget to lock up, but here they were. Well, he and Pietro were good friends. He put some money by the register, then stepped behind the counter. Kayleigh would get a kick out of this! He leaned over the counter, smiled, and looked into her eyes as he asked, "What flavor, miss?"

She giggled and replied, "nilla!"

"Vanilla it is, then!"

He heard something faint from the backroom. He paused and listened.

Nothing…I must be imagining things. This lockdown is getting to me.

With a shrug, he reached for a small cone, then turned it slowly under the soft serve dispenser as he pulled the lever. A little sloppy at first, but he got the motion right pretty quickly. He released the lever and turned to hand the cone to a brightly smiling toddler when he heard another noise. It sounded like a whimper.

Dario felt a chill of fear, then started walking to the end of the counter. A portal separated the retail area from the back room, where Pietro had a desk, storage areas, and so forth.

"Pietro, are you here?"

Silence. *Weird!*

He shrugged. Must be his imagination! He glanced back. Kayleigh was licking her cone, ice cream smeared all around her mouth and all over her hand, legs kicking to and fro as she sat, a smile that would win any heart on her little face…

A long, low, guttural moan sent chills up and down his spine.

"Pietro?"

Dario approached the portal. Unlocked, it cleared. As it cleared, he stared. Pietro, naked, moaning, was in the maw of a nightmare. Claws held his chest, and his legs were mid-thigh deep in the kraken's beak. Tongues slid fluidly across his torso, shoulders, and neck. The tongues extended gently, lovingly, caressing his flesh, then tensed and pressed and drew blood. Two of the kraken's six eyes turned from Pietro's eyes to look at Dario as Pietro, turned his head and rolled his eyes to look, pleadingly, at Dario. He whispered hoarsely, "Help me. Please, help me. It burns! Oh, God, it burns! Please, kill me…"

The tongues tensed and Pietro's eyes returned to stare into the krakens. Blood seeped from where they pressed against Pietro's skin as he slowly slid more deeply into the kraken. A brief shriek ripped from him as the kraken's flanks flashed bright colors faster and faster.

Dario stepped backward, thinking of Kayleigh, but he found himself unable to move. He heard Kayleigh singing one of her little songs from nursery school; he heard random noises as she apparently wandered about the store. He knew, somehow, that the thing realized Kayleigh was nearby. The colors along its flanks flashed even more brightly and rapidly as two of its eyes returned to Dario.

Pietro shuddered as he slid more rapidly into the thing's gullet. As his waist disappeared, his back arched and he began screaming, "No, No, NO!"

"Daddy? What's happening, Daddy? Is Pietro playing a game?"

The kraken stretched its tongues, pressing them against Pietro's shoulders, and pulled him into itself. He was still screaming as his head disappeared. It turned towards Dario, reaching out with a claw to push him aside.

It went through the portal, and Dario's heart broke as he heard Kayleigh scream.

▪ ▪ ▪

The lovestruck teenagers held hands as they stepped through the portal. They giggled as they walked along the path towards a private place.

Her parents didn't approve of him; and, at some level, he knew that was part of his appeal to her. But he didn't care. They were adolescents. Though the two professed their undying love to each other and to anyone who didn't manage to walk away quickly enough, even they knew, at some level, that this was hormones and rebellion and experimentation and wouldn't last. But, it brought pleasure and smug satisfaction and filled them with an infectious, cavalier recklessness. The blinded lovebirds would have been mortified to learn their pragmatic parents understood all too well, assured each other they'd outgrow it, and, privately, shook their heads at the memories it invoked.

They were supposed to be in their quarters with their respective families. For their own safety, the two were assured. But, the temptation to slip away was irresistible. Like enchanted children, they walked the pathway encircling the residential structure, held hands, leaned into each other giggling, and ignored the storms outside the dome. They walked between trees and bushes until they came upon a thicket, a shrubbery cluster surrounding an open space containing a soft mat donated who knows by whom or how long ago.

A quiet little retreat discovered many times over many years by many adolescents.

With a few glances confirming no one was looking (and a wee bit of disappointment that no one was looking), after a few giggles, the two stepped into their place and tore away each other's clothing as they kissed.

Neither one noticed they were not alone. Not until they stopped moving. Despite their mutual lust, they stood, in very awkward positions, staring at each other with confusion.

Then, part of the shrubbery moved.

As claws slowly extended, the kraken's coloring and texture changed. Its skin smoothed, and green turned to accelerating bands of colors flashing along its flanks. He stared at her as she was lifted effortlessly by the claws, her remaining clothing torn away by the kraken's tongues that then pulled her feet into its beak, pressing them into itself as four spindly fingered eyes stared into hers. The other two stared at him.

Unable to move, chilled to the bone with fear, tears flowed down his cheeks as he watched her moan in agony as she was slowly drawn into the kraken's mouth. As she was gradually consumed, as the tongues explored her body, they probed her genitals. And, as he stared, he realized he was becoming aroused.

Suddenly, he was painfully aware of her; he realized he was being controlled. He tore his eyes from her and stared into the two eyes focused on him. He felt his emotions being manipulated. He was cold, then hungry, then thirsty, then tired, then anxious, then aroused once more. As his arousal increased, he stared at her once more. Her genitals were disappearing into the kraken's beak; he could not look away and it was wrong. Just *wrong* to be feeling what he was feeling but it wasn't up to him.

He climaxed.

As his body relaxed, he realized the kraken was playing with him, experimenting with his emotions and its ability to control his body and mind. And he was terrified.

He watched the tongues explore her upper body and caress her breasts; then it drew blood from her breasts, and he was not moved. Instead, he began to feel what she was feeling. He felt her fear and her agony and he knew, as she knew, that the thing was slowly digesting her. He felt the tongues exploring *his* body as if they were moving across his flesh instead of hers.

Then that feeling was gone.

He felt nothing but his own terror and remorse as he stared at her. He watched her eyes roll back and heard her last, awful scream. The thing's tongues reached out, pressed against her, and pulled. He watched the crown of her head disappear inside it. He regained some motor control.

He managed to shake his head, side to side, to softly plead for his life as it reached out for him. He knew that there was no hope, there was no escape. With her gone, there would only be an eternity of pain as it consumed him as well, but far more slowly.

DAY 15
ESCALATION . .

Pavel stood at his console looking at a wireframe display of the primary dome. His peers in the room kept him appraised of their progress, fed him information as it developed, all against a buzzing background as they engaged with their respective teams. He listened to reports from his squads and then gave them orders. He was stressed, of course. Came with the job. Wasn't his first crisis, though it was the biggest he'd managed. He hoped to never face anything like this again.

Mindful, deep breathing helped, as did periodically remembering to unclench his fists.

Another report came in. A farmer spotted a kraken taking one of his goats. He reached for the cluster of wireframes hovering to his left, pulled the highlighted agricultural dome and positioned it over his console. As it expanded, the primary dome display shrank and flew to the hovering cluster. This was the first reported livestock loss, and that meant the damn things weren't confined to the primary and residential domes.

Not surprising, of course. He'd expected it. But the confirmation was anything but welcome. His options were lousy. He could reassign squads from the residential and/or primary dome, which would almost certainly cost a lot of lives; or he could ignore it and let the farmers fend for themselves. Any loss of life was unacceptable, but there were a lot more people under the primary and residential domes than any of the agricultural domes. Of course, livestock losses were a whole other issue, and kraken fattening on livestock weren't good for the colony, either.

Damn…

If farmers had the means to slaughter livestock, then they had the means to take out kraken. He prepared a note suggesting the farmers take up whatever tools they could use as weapons and hunt kraken before they grew big enough to threaten farmers. Pavel suggested they start by hacking off their sensory clusters. That's what one of the xenobiologists reported!

He sent the note, then sent an update to Yoshi and the rest of the team. He also sent a request to Jack about the possibility of sending some weapons to the aggie domes. He might not be able to spare any squads, but they might be able to spare some weaponry. Jack came back with a suggestion from Eden colony. He was considering diverting some of their fabrication capacity to armed drones. Got around the weapons training constraint and would do well under the aggie domes. Yoshi supported the idea if Pavel thought he could spare the weapons and ammunition constraints for his squads. They had a quick exchange about exactly what that constraint would amount to, then Pavel encouraged him and asked for updates as that progressed.

Then another report came in.

Pavel reached out and pulled the residential dome wireframe back to his console. As it expanded, it color coded his areas of interest: ductwork and infrastructure, where the crawlers were, where the sniffers reported kraken scent, where pets had been taken.

And three red stars.

One was within the arc of the residential structure; the other two were along the path between the structure and the dome stem wall. He hesitated, then swore. Pavel stepped back, and, rather than send a message, called out.

"Yoshi! We're getting reports of kraken attacking colonists!"

The room went silent.

Yoshi froze for a moment. "Okay. . . People, it's started. We've been expecting this, so let's remain calm and work the problem. Pavel, get your squads moving!"

▪ ▪ ▪

Amy and her squad were the closest.

They were patrolling the center section, the original section, of the residential structure. Both their pet disappearance modeling and Kim's reasoning suggested they'd most likely find kraken here, and both were right. Over the last three hours, they'd dispatched four of them, two while they were consuming dogs. They were bigger than she'd expected. One had to be 35 kilos.

They were on the second floor, standing by the smoldering remains of a kraken that had just feasted on a cat when she got the call. Amy hesitated. It didn't seem real, the whole thing. It just couldn't be real! But it was.

"Kraken attack! Downstairs, by one of the shops. Let's go!"

They followed her through the corridor, bypassing the lift to run down the stairs. They went through a portal and paused. She checked her wrist computer to see where the attack was relative to their position, then turned left. They walked across a grass field between shade trees. The dome maintained a very comfortable 22 degrees Celcius, but the midday lighting was intentionally intense, making the shade enjoyable for residents taking breaks.

Amy and her squad ignored the shade, fanning out as they approached the ice cream parlor. The sniffer attached to her flamethrower wand definitely picked up kraken scent, but the motion sensor beside it recorded nothing but them. At her direction, two went to the left, and they met in front. Nothing yet.

But their sniffers were clear; there was a kraken nearby.

Jim and Naomi stayed at the door while she entered with Stu and Raul to find overturned furniture, debris on the floor, and a disturbingly large kraken dropping. She scanned it, picking up three wrist computers, including one belonging to a child.

"Dammit!"

A quick walk through convinced them the kraken wasn't inside. They rejoined Jim and Naomi, then fanned out, walking away from the building, looking for the kraken.

■ ■ ■

There were too many, and that was unfortunate. They were like the host! Delightful, delicious and so emotive, it was so pleasurable to feel their pain and fear as they were consumed. The kraken was relieved to abandon the dark spaces. In the dark spaces, it felt safe. And there was food, though nothing as satisfying as the host. Still, it managed to grow, feasting on the prey that roamed the dark spaces until its own growth rendered those things dissatisfying.

The kraken reached out with its senses, exploring the compartments adjacent to the dark spaces. There were prey, like the host, but too big and never alone.

It found other prey, though. The kraken measured the sharp teeth and claws of the residential cats and dogs. The dogs, in both size and shape, were initially confusing, but electromagnetically, they were the same kind of prey. It could not wait to taste them!

The kraken studied them, then experimented with suppressing them, then broke through the barrier from the dark spaces and feasted on them. They were tasty and emotive; it enjoyed those feedings very much. But nothing compared to the host. It continued to grow; as it grew, it was emboldened. It stopped returning to the dark spaces and tore through the barrier separating the compartment from the wider space beyond.

The wider space was brightly lit, which made it anxious, but the space was easier to move through. The kraken found more compartments holding all sorts of prey. Hairy ones, but also other kinds usually bound in various containers it could easily break through. It grew and grew and grew, until it was capable of taking one of the prey like the host. If only it could find just one or two alone.

Then, the kraken smelled death.

Alarm pheromones were in the air. Alarm pheromones suggested something was also roaming these spaces. Something was killing its kind! The kraken wandered more urgently, moving away from the smell of death; eventually, it found itself in an even broader space. There were big, green things all around, things that were easier to match and hide. Its confidence returned as it roamed, reaching out with its senses, until it found a structure with *only one prey* inside!

It entered the structure, bombarding the prey, then took it. So amazing! So delicious! So emotive! Feasting was such rapture!

Then, as the feasting drew to a close, two more approached! And one was small!

The kraken suppressed the larger prey as it approached, then finished ingesting the quarry it had been enjoying. Trembling with anticipation, it moved towards the little one, anxious to clear away the annoying encumbrances and taste its flesh. But then, it picked up on something new. It sensed the prey's anxiety with respect to the smaller one.

It paused, then decided.

It moved past the new one and took the small one first. It backed up so that the big one could watch as it cleaned the small one, then consumed it slowly. The small one made all sorts of exquisite sounds as it was consumed, as its flesh was rendered, as its extremities were digested, and, all the time, the kraken could feel waves of pain from the larger one!

As the smaller one expired, as the kraken ingested the last morsels, it decided it would have to recreate this delightful experience. After the larger one was consumed, the kraken waited a while, hoping more prey to approach. But none did. The kraken left the structure and moved cautiously across a field of green when it detected approaching prey. Quickly, it realized there were too many; it moved towards a larger green thing, matched its colors and textures, and stood absolutely still.

▪ ▪ ▪

About 10 meters from the ice cream parlor, near a large hyacinth shrub, Amy thought Jim and Raul slowed. Raul raised his hand, shook his head, and said, "whoa!"

Jim and Raul stopped moving as Amy, Stu, and Naomi encircled the shrub. Naomi froze as well. Amy felt a little fuzzy, but not Stu. He raised his flamethrower and doused the shrub.

As he did so, part of the shrub backed away. The green leaves and bright yellow flowers disappeared, revealing a very large kraken, easily 50 kilos. It backed away, slowly. The squad, no longer impaired by the thing, formed a circle around it, dispatching it quickly.

"That was wild!"

Amy glanced at Raul and nodded. "You ok?"

"Little bit of a headache, but it's clearing."

"Jim? Naomi?"

They looked at each other and shrugged. "Pretty much the same," Naomi added. "I've read they can cause brain damage after a bit. I can believe it! But I don't think it lasted long enough."

Amy cocked her head, then said, "You keep me posted how you're feeling. If that headache doesn't clear, you're getting checked out."

She met their eyes one at a time so they'd know she meant it, then sent in her report. The monsters were growing *fast*, and this one managed to suppress three of them. She advised Pavel they'd need to combine squads soon, before the things could take five humans at a time.

DAY 16
COURTSHIP

"I need a break!"

Pavel stepped back from his console, closed his eyes, lowered his chin, and punched his nose as one of his leads took his place. Yoshi and Jane walked over and escorted him to a chair. Jane asked if she could get him anything.

He smiled ruefully and joked, "more hours in the day? Make me ten years younger?" He waved his hand weakly and asked, "Would you please bring me some water? We'll start with that."

Yoshi, taking a seat beside him, asked, "How are we doing?"

Jane returned with his water. She also brought a nutrition bar and a cup of coffee. He thanked her, sipped the water, smelled the coffee, then unwrapped the bar and took a bite as Jack, Keith, and Kim joined them.

Pavel chewed, then sighed. "Well, let's see. Here at the primary dome, we've destroyed about a hundred kraken. Just small ones, all picked off by the crawlers. I'm not suggesting there aren't big ones over here, but nobody's seen one yet that's big enough to pose a threat to colonists." He shrugged and looked up before taking another bite. "No idea why. Maybe they exhausted the vermin here, then went over there looking for more. No idea."

"That's a reasonable theory," Kim volunteered. "Sure, we have vermin everywhere, but the residential dome population is much higher than the primary. We know, from our traps, that the vermin population is always higher there. Bigger critters, and more diverse, too."

Pavel shrugged. "Makes sense…Anyway, that's the good news, right? Until and unless we see something bigger over here, I've reassigned all the squads to the residential dome, because things aren't going so well over there. We have reports of kraken affecting three or four squad members at a time, so we merged squads for teams of ten. Slows us down, sure, but we can't afford to lose anybody."

He straightened in his chair, putting elbows on the table. "We've destroyed sixty or so at the residential dome. The teams are finding them outside, and sometimes it's pretty clear they've taken colonists that either hadn't locked down yet or didn't believe they had to or whatever damn fool thing they were thinking. God knows. But all that's left is torn, bloody clothing or a pile of kraken dung with wrist computers in it."

He sighed again, shook his head as he looked at the floor. He took another sip of coffee and continued. "We're also finding empty quarters. In some cases, the portal wasn't secured. Sometimes they tore through a corridor panel; at other scenes, it looks like the kraken went into the plenum and over the wall. These residences really aren't that securable, y'know? Never thought they needed to be…No idea how many are gone, no idea how many people were in each residence. Sometimes you can make an educated guess from the debris."

Pavel took another bite and chewed. "We're seeing them elsewhere, of course. We're investigating a couple of reports from municipal infrastructure, the agricultural domes, about taken livestock. Farmers are doing what they can to protect their animals. But we can't send any help, not yet. I can't divert a squad from the residential dome. I just can't!"

Yoshi nodded. They all knew hard decisions would need to be made, and he remained confident that Pavel was making the best, or perhaps the "least bad" decisions possible. He turned to Jack and asked, "How are the electromagnetic gizmos coming along?"

Jack shook his head. "They require components we don't normally stock, at least not here. Eden colony had the same problem. They had to cannibalize some of the orbiting platform components to make them, which limited how many they could make until they could fabricate said components. The stuff we *do* stock is up the stalk. The orbiting platform spaceport has them for spaceship maintenance and refits, y'know?"

"So send some down!" Jane interjected.

Jack shook his head again. "Even at top speed, it takes a single gondola days to descend the stalk. Landing craft are faster, of course, but we'd need multiple trips. No, we can just fabricate the parts down here faster than we could send them down, but it still takes time. We expect to have some ready tonight." He nodded towards Kim and added, "Kim's people are already training with the few we completed. They don't require the kind of precautions weapons do, but they do require some proficiency. Anyway, we think that, as we make them, we'll add 'em to the squads with people ready to use them. Sound good?"

Everyone turned to Pavel, who nodded and held up a thumb.

Yoshi asked, "how about the drones?"

"Similar problem. We don't normally stock drones, so we're building them from scratch. Eden had plenty of drones. They use them for all sorts of field work. We don't! But most of the components are standard stuff, thank God. On the other hand, we don't normally stock explosives, so we're also making the charges. Eden colony sent us what they were using against the pineapple things. It's basically a projectile, a big bullet, for drones to fire at big kraken. On impact, a small charge fires shrapnel into the body. Has to hit them between ribs to do any real damage." He paused and shrugged, then continued, "Keith and his Eden buddy George are helping us understand how tough these things are. They're practically armored!"

Pavel, nodding, added, "It's been hard on the farmers. We've been as clear as we could that they should only try to take on kraken that are still small. Don't want to lose any farmers! But all they have's agricultural equipment, so their best bet is to try to lop off sensory clusters. We've told 'em anything bigger than a dog would be a threat to them, but I'm sure it's hard watching one of these things taking your livestock. Can't know how many farmers we've lost trying to take on bigger kraken." He turned to Jack. "We *need* those drones!"

Jack nodded fervently. "You'll get them, as soon as we can make them."

Pavel's shoulders sagged as he closed his eyes and nodded. "I'm sorry, Jack. I know you're working as fast as you can. I didn't mean to imply anybody was slacking off. It's just that…"

Jack placed a hand on Pavel's shoulder. "I know. People are dying." He shook his head, then muttered, "We should *not* be dealing with this bullshit! Damn that ship!"

Yoshi straightened and retorted, "enough of that! Lubandi and her people had no way to intercept that ship, just like we have no way to reach the *Eureka*."

"They could have warned us!"

Yoshi looked at him, then added, "They did. I checked the logs. Fifteen years ago, they sent my predecessor Kelty a warning, just like we'll send a warning to *Eureka*'s destination. But it was fifteen years ago, and none of us were here back then. None of us knew." His voice softened a bit as he added, "We can talk about protocols after we get through this to make sure such warnings aren't lost."

After a few moments of painful silence, Keith raised a finger before contributing. "I've been thinking about your squad sizes. The kraken don't seem willing or able to take on larger groups. Maybe we need to rethink our lockdown strategy. Maybe we shouldn't have small groups. Maybe smaller groups should coordinate so kraken pass them by!"

Yoshi nodded. "That makes sense, but our primary job is killing as many kraken as we can, as fast as we can, so we can't task squads with facilitating that. And I don't want unarmed personnel wandering around trying to move colonists. Doesn't make sense to expose one set of people trying to save another set. But we can put out an advisory, let smaller groups sort it out for themselves." He turned to Kim and asked, "How can we most effectively communicate this with the squads? Put up an app that'll let folks enter how many are clustered together, match 'em up, guide them to combine groups where there's enough room?"

Kim nodded and gave a lighthearted smile. "We're on it!"

▪ ▪ ▪

The kraken moved confidently through the corridor. The smell of death lingered in the air, but it no longer cared. It was big enough now that it feared nothing. As it progressed, it picked up the electromagnetic signals of prey. It zeroed in on a number of them congregated together, in one of the compartments.

The kraken followed the signal, then paused. Gradually, steadily, it bombarded the humans it sensed on the other side of the barrier. As it did so, some, at least, became aware that they were being bombarded. But they also knew they were helpless to resist. The kraken could feel their fear building.

So delightful!

The kraken picked one that was particularly malleable. It did not understand how the portals worked; it only understood the prey had the means to open them. It pushed its dominance, harder and harder, until the prey touched the portal. As the portal cleared, the kraken squeezed through. It scanned the room. Some were standing, others sitting. It allowed them to move their heads, and they turned to look at it. Waves of intense fear radiated!

Just delightful.

Some just stared; others secreted moisture from their eyes. One lost control of its bladder, and the smell of its nitrogenous waste blended wonderfully with the fear. As the odor wafted through the room, another released its nitrogenous waste as well. The kraken remained still, enjoying the fear.

But it was *hungry*.

It decided to consume one quickly to ease its hunger. That would also heighten the anxiety of the others. The closest one, the one that had let it enter, was twitching. The kraken had pushed this one hard, damaging it in the process. It wasn't generating as much fear as the others anyway, so the kraken decided to eat it first.

The kraken stared into the prey's eyes as the human slowly realized what was happening. Consuming it quickly detracted from the enjoyment of its pain and fear, but such speed did provide an explosion of flavor as the prey's lower limbs dissolved. The kraken's hunger abated, it paused to consider the remaining prey. They were *so* emotive! Yes, delicious, and such *fun*!

It could feel their anxiety as it pondered which to take next. The kraken trembled from pleasure as it felt their fear and horror. One by one, it selected its victims, ceasing their suppression; their struggling exertions only intensified the kraken's pleasure. The food began to shriek, louder and louder, and the kraken trembled with pleasure.

It feasted, one by one. It experimented with different techniques, with different levels of suppression and stimulation. It tried different ways to share the experience of the individual being consumed with the other prey so they could *all* share the experience. The kraken was barely able to contain itself. The prey were so delicious, emotive, and, fascinating.

There were only two left when it smelled something *new*.

Something wonderful. Something unlike anything it had ever smelled before. Something it simply could not resist! The kraken remained motionless, moved profoundly by the smell, but not knowing what it needed to do about it.

Then, suddenly, it *knew*.

The kraken looked at the remaining prey. It decided. It selected the smaller one, stripped it, and pressed it headfirst into its beak. Sustenance, nothing more. It reached out for the last one.

It had a special use for that one.

Gently, but firmly, the kraken stripped the final prey and carried it through the portal. It had grown while feasting, enough to damage the sides of the portal as it left the compartment, injuring itself slightly in the process. It didn't care. It knew it would heal. All it knew was it had to find the source of that new smell, *right now!*

DAY 16

PROGRESS

Elke walked into her conference room to find George staring out a window at the Garcia Mountains. As she sat, without turning, George made an observation.

"I sometimes wonder how Sam would have felt having those mountains named for him. They're beautiful, of course, but, for an evolutionary biologist like Sam, the biodiversity is just stunning."

George turned and walked towards the table. Sam Garcia was Eden's first casualty, attacked by a mysterious marine creature during a technical dive. His remains were never found.

"We've named a number of species for Sam. It's funny. I can picture him arguing against that, arguing the organisms should instead be named something descriptive! But, whenever we discover a species we think he would have found intriguing, we name it in his honor. Which, I suppose, means such names are at least somewhat descriptive!"

Elke smiled. She didn't know Sam as well as George did, but she'd known him well enough to believe George was right.

"Tell me what's going on."

George kept his eyes on the mountains as he explained. "I've been working with Keith Saputra, the xenobiologist you referred me to from Alexandria's Deep Space Academy. He's a biochemist by training, which makes sense. The planet's rich with novel organic chemicals, but no life forms, not even prokaryotes, uh, bacterial equivalents. He's done some very important work there within his specialty! But, Keith's an administrator, so he does have oversight of their academy's agricultural programs. They recently built a zoological/botanical park. Anyway, he was the best local choice for us to work with."

George shrugged, then continued. "We just received an out of cycle update. Keith says they're picking up sex pheromones. Things are about to get a whole lot worse for them."

Elke closed her eyes for a moment as her mind flooded with memories. Kraken large enough to emit sex pheromones would be bolder, reckless even. They'd start choosing colonists as gifts. As George once explained to her, if you're a cannibal, offer your prospective partner a meal to occupy it while you're mating. Kraken are also hermaphrodites, so any two kraken could impregnate each other. Once impregnated, they'd be even more aggressive. Failure to maintain enough nutrients in their blood stream would lead to cannibalization among their unborn offspring.

Elke opened her eyes and asked, "What can we do for them?"

"Not much, unfortunately. They're making solid progress. In this morning's update, Keith reported that, under their primary dome, they're killing small kraken by the dozen with their duct crawlers, but they're not seeing any large enough to threaten colonists. They're working theory (and this makes sense to me) is that they spread through the colony while they were smaller, and the growth rates of those that remained under the primary dome has been constrained by a smaller vermin population. It just makes sense that there'd be more vermin under the residential dome, so those kraken have grown a lot faster. Then, they moved on to pets, then people.

Anyway, he says they've reallocated all their squads to their residential dome. They're getting reports from agricultural and infrastructure domes, but they don't have enough people to send them any help. Farmers are doing their best to defend their livestock. I don't like them having to make that choice, but he's right. They need to focus on killing as many kraken as they can before they breed."

Elke, no stranger to tough calls, nodded.

"Keith says they've killed at least a hundred kraken that were over 20 kilos, and at least half of those were over 50 kilos. No idea what their casualties are. Of course there are always people that refuse to obey lockdown orders, but none have been observed so they're likely to be lost. And they're finding a lot of empty residences. No idea how many colonists that translates to. But he says they're guessing they've lost hundreds of people already."

Elke turned to look out the window and nodded.

"We sent them specifications for drones armed with flechettes," George continued. "We didn't use them for our kraken outbreak since we were aboard the orbiting platform. But they were very effective against the pineapples, so we think they'll be useful to Alexandria. They don't have any drones, you know? Not much use for them on their planet's surface. They've been able to fabricate drones and should have some up today, which will help them spot mating kraken. It's taking them a little longer to make the explosives for the flechettes charges. They don't need explosives at the established settlement, and it was easier to make some from scratch than to try to bring any in from the construction of the new settlement, but it still takes time. Anyway, they should have drones firing charges against the kraken later today or, worst case, tomorrow. That'll be a game changer!"

"What about electromagnetic interference tech?"

"They've built some and are distributing to the squads. We should hear something back about their efficacy soon."

Elke nodded, then said Nikko was sending them everything they had about dealing with spores. Eden's kraken outbreak had been different. They lost a lot of people to kraken; but, by the time the kraken were ready to produce dragons, they'd killed all but one. That kraken's dragon offspring certainly caused their share of damage, but over all, Elke and her crew had been very, very lucky when it came to minimizing the release of spores.

Elke wasn't one to depend on luck. Their kraken research included coming up with strategies for any future encounters, including dealing with spores. They would make sure to keep Alexandria informed.

"Nikko sent them spore filter specifications they can add to their life support systems and scrubbers they can add to the drones. Let's hope it doesn't get that far, but, sad to say, I suspect it will."

George nodded, his eyes glassy. He never took his gaze off of the mountains. Elke couldn't tell if he was still hearing her.

"You know, George, this is going to get bad. Yoshi told me the *Eureka* might, just might, be carrying someone that's infected. We need to prepare their destination. They're doing a twelve year crossing to one of the older, much larger colonies."

George snorted. "Yeah, well, you warned Alexandria and look what good that did."

She raised her eyebrows. "We sent that warning fifteen years ago. I remember explaining it all to Avram Kelty, but he filed it and then signed on for a crossing. Gulati came in and never saw it. I should have realized, should have refreshed the warning as the *Lucky Strike* was approaching. This is on me, George. All on me!"

He looked her in the eyes. His voice was kind, but his eyes were tired, even sad.

"Elke, we've known each other a long time. Don't do this to yourself. It's not your fault."

There was something in his expression she just couldn't read.

She let it go, then smiled. "You're as bad as John!" She sat upright and rolled her shoulders.

"Well, George, as I feared, I'm seeing more chatter about the dual hub proposition, including a broad range of ideas to eradicate the kraken. I'm starting to see those discussions include pineapples as well." She shook her head slowly, then added, softly, "Can you imagine if those idiots managed to get pineapples offworld?"

DAY 17

BREEDING

Jack jumped back from his console and cried out, "We have kraken mating!"

As the others rushed towards the console, Jack reached for one of the hovering projections and dragged it to the conference room table. As it expanded, they could see two large kraken on the path between the residential structure and the dome stem wall. They were lined up, head to tail, each reaching across with the middle dorsal limb to enclose the tip of the third dorsal limb of their partner. Each was slowly consuming a colonist clutched with their claws, drawn feet first into their beaks as they writhed and screamed.

Keith pointed at the image. "Remember, they're hermaphrodites, so they impregnate each other. Those limbs that were folded in sheaths in their backs are their sex organs. The male sex organ cups over the partner's female organ, delivering their equivalent of sperm..."

"Can they be saved??" Yoshi, more concerned with the suffering colonists, interrupted to ask Jane.

Jane stood and responded. "They've lost their legs by now, at least from the knees down. If a squad can recover them *right now*, we should be able to regenerate their legs. I imagine there's substantial damage to their brain tissue as well, which should also be repairable."

Pavel shook his head. "Nearest squad is a good twenty minutes away. I can't get them there in time!"

Keith, chastened, jumped back in. "By the time they get there, the kraken will have finished consuming the colonists, uncoupled, and resumed hunting. We need to focus on killing them while we can!"

Yoshi nodded, then turned to Jack. "The drone that's taking this image, is it armed?"

Jack nodded.

"Do it!"

Jack swallowed hard, then tapped his wrist computer. The image shifted subtly as the drone descended, managing its focal length to keep the image size consistent. The kraken, rapturous, ignored the drone. Two puffs could be seen from the drone's point of view as flechette charges were fired. Both kraken shifted eyes towards the drone.

The first projectile hit the kraken closer to the drone between the claw and first walking leg of its left side; the second projectile hit the other kraken at the base of its male organ. As soon as the projectiles touched the krakens' hides, they exploded; casings popped off the targets as their explosive driven flechettes penetrated the krakens' bodies. The kraken's bodies expanded dramatically, then sagged. Their claws remained locked on their victim's bodies, but their tongues stopped moving. Their legs gave way, and their bodies hit the ground. Their sensory clusters fell, landing on their beaks, with their eyestalks sprawled.

Neither kraken moved.

The room remained silent and stared at the two colonists, still held in the kraken's claws, hanging limply from the krakens' beaks.

"What happens now?" Jack asked, his voice quivering with fear and concern.

Jane shook her head. "I…I don't know. Pavel, I don't know if they'll still be alive when your squad gets there."

He looked at her, then responded. "If I send a squad, what then? They're not medics. They can't save them. And we can't afford to redirect them from killing kraken to rescuing colonists! Not if we're going to keep the damn things from producing offspring. Can you get a medical team there?"

Jane sighed. "Probably not in time, and if you can't spare people to escort them…" She shook her head again, then stepped away.

Yoshi looked at Pavel. "Can you euthanize them?"

Pavel swallowed hard, then said, "I can hit them with flechettes. That would kill them, but not gently, and not quickly, if that's what you're asking."

"That's what I'm asking."

Pavel stared at them as they struggled to free themselves from the claws. "I thought the kraken were dead. Why are they still holding those poor bastards?"

"They're dead, but kraken claws lock," Keith responded. "It's actually an interesting structure, but that's not important right now. Forcing the claws open wouldn't be easy, and you'd need some tools."

Jane returned to the table. "I have a couple of doctors and nurses on their way. They don't live far, so they'll take the victims to the nearest medical office and try to stabilize them."

Pavel looked at her, then shook his head. "Are you sure you want them to do that? What if they're attacked?"

Jane did not take her eyes from the projection.

"They volunteered..."

▪ ▪ ▪

The kraken was torn. The scents were still enticing, but they were no longer irresistible. After repeated matings, all immensely pleasurable and satisfying, it no longer felt compelled. But it remained hungry. In fact, deep within itself, it felt a shift. Something was happening; something it did not understand. But, somehow, it knew its hunger would soon increase.

Like Earthly forms, the creature's body was regulated by signals carried by specialized cells strongly resembling neurons and chemical signals similar to hormones. A hormonal shift, triggered by the reception of sex pheromones, drove the creatures to seek each other out to mate. And, with each mating, a different set of hormones spread through their bodies, driving additional changes. As these hormones accumulated, their interest in additional matings eased. Then, their reproductive systems shifted, for it was time to sift through the genetic material received from its partners with its own, to fertilize its eggs, to form zygotes, and to encourage their implantation in its womb.

As their reproductive systems shifted, those very systems acted almost like an analog computer. Various tissues and organs of their bodies, in response to stimuli like injuries, fear, alarm pheromones, hunger, and a host of other experiences, secreted hormones that bathed their reproductive systems in chemical messages. These messages, accumulated through their life experience, drove each kraken's "decision" regarding the timing and prioritization of gamete fusion and the secretions within its womb that would govern the development of resulting zygotes.

In one, a large number of zygotes, as diverse as they could be within the constraints of consistent genetic material, implanted and formed independent connections with their parent's body (analogous to an Earth mammal's placenta) and began to grow. In another, a few diverse zygotes implanted. They each split a number of times, forming placenta-sharing clusters of identical zygotes, then initiated their development. Eventually, they would emerge as small, tightly bound groups altruistically committed to the release of spores by at least one group member. In yet another, one zygote implanted. It split into two, establishing separate connections to their shared implantation point. One, the alpha, began to develop. As it did so, it produced hormones that stimulated its placental connection to prioritize its nutrition and bathe it in royal hormones. The other split again.

Of the two, one split into two identical zygotes. They formed their own links to the placenta, prioritizing their nutrition (though not above the alpha) and began to grow. They would be the elites, altruistically committed to each other and the alpha. They were ready, if the alpha were lost, to step up as the alpha's successor.

The other zygote split again, and again, and again, and then again.

Fourteen formed placental links and began to grow. Specific genes activated that would result in the morphological distinctions of soldiers. The remaining two split again, and again, and again, and again, to become thirty-two pawns. Their placental links saw to their nutrition only after the alpha and elites and soldiers were fed. When their parent's hunting faltered and nutrition levels sagged, the alpha and the elites and the soldiers would, as needed, consume their pawn wombmates first, though not lightly. The alpha would lead the pack in their shared goal of seeing to the release of spores by the alpha, elites, soldiers, and as many pawns as possible; but, pawns would not be spent capriciously, not in the womb or after. But, they would sacrifice themselves willingly to ensure their reproductive success through as many of their siblings as possible.

Each creature's body chose a reproductive strategy, placing an evolutionary bet based on its life experiences within the rules of the evolutionary game played by countless generations of ancestors. And, as each creature chose its strategy and placed its bets, the initiation of their young's development drove their hunger to new, unprecedented levels.

▪ ▪ ▪

Amy and Chad lead their combined squad along a broad avenue connecting the main entrance of a residential structure to the dome's transportation hub at the base of the central tower. Normally, it'd be a very pleasant walk. Greenery, grasses, flowers, shrubs, and shade trees lined the boulevard, periodically interrupted by playgrounds, historical markers, and a variety of commercial operations ranging from novelties to food and drink. Normally, this time of day, they'd make small talk as they shared the boulevard with hordes of pedestrians and the occasional vehicle.

But this was not a normal day.

They'd just incinerated a particularly large kraken on the third floor of a newer section of the residential structure. There were residences behind it, and one could only guess which of those residences had been raided by the kraken. They decided not to look. They would have checked yesterday. They would have gone door to door, relieved if they found unharmed residents, but accepting if they found the empty aftermath of attacks.

They'd been told their primary mission was to kill kraken, not rescue residents.

They'd been told that attempting to rescue residents would actually result in more losses.

And all that made rational sense, but it just didn't feel right. So, Amy, Chad, and those under their command ignored instructions and looked for survivors to rescue. But, there weren't any. When the kraken attacked, everyone disappeared. *Everyone...*

It quickly became overwhelmingly depressing, so they stopped. This undeniable sorrow and anger motivated them to hunt down and destroy every kraken they could. It was hard to step away to stroll towards the transportation hub, but orders were orders. They trusted that those giving the orders knew something they didn't.

As the newly formed squad approached the tower, a kraken rushed towards them. They'd never seen one moving like this before. It wasn't stalking them. It wasn't camouflaged at all. Instead, bands of yellows and reds flashed along its flanks. It seemed to be running away from something more than running towards them. Amy and Chad exchanged a quick glance, gestured towards their squads, and unslung their flamethrowers.

The kraken didn't slow, creating the impression it planned to run through their group. But, as soon as it was within range, they hit it. The creature slowed, then stopped, then collapsed in a smoldering heap.

Chad shook his head. "Well, that was new!"

Amy, looking over his shoulder, pointed towards two approaching women accompanying a suspensor palette supporting an apparently electronic device. It was about a half meter square, with air vents and a variety of displays and control interfaces.

Once they were close enough, the taller woman said, “Looks like we flushed one for you!”

Chad, looking at the box, asked, “Is that what I think it is?”

The shorter one replied, “an electromagnetic disruptor? Then yes!”

Once introductions were made, the taller woman, Maeve, explained the device could both detect kraken electromagnetic signals and generate a field that kraken found unbearable. The shorter one, Poorna, added they’d also launched the first set of drones, which were now patrolling the dome.

Amy nodded. She looked at Chad with hope. “Maybe we’re going to make it through this after all…”

DAY 17

ESCALATING HUNGER

Its young were increasingly restless.

They were growing rapidly, drawing sustenance from the parent kraken's body. It knew that, if it failed to maintain adequate nutrition levels in its body, the offspring would turn on each other. In fact, they were very close to doing so! So, it hunted, desperately and relentlessly, consuming one prey after another. No time to enjoy them, no time to share the experience of their consumption with them. *Sad!* But there was no time! It walked through a corridor, reaching out with its senses.

There!

It felt the electromagnetic signatures of prey! There were three of them, wait, no, four! It suppressed them quickly and harshly, then tore through the corridor wall to enter their quarters. A small one was horizontal, its body spasming. This happened with such intense suppression, but the parent didn't care. The kraken reached out with its claws, pulled the small one towards it, and used its tongues to strip the annoying inanimate material from its body. It didn't bother to investigate the prey's body. It reached out with its tongues and pulled the prey rapidly into its beak. It felt the prey die as digestive juices went to work; it could feel its young slow their movements as the nutrient levels in its bloodstream improved.

Its young were still restless, though.

The kraken could feel the other prey looking at it; the kraken could feel their fear and anguish as they watched their young be consumed. It reached out for the next one and tore away its coverings; it felt the prey's fear as it pulled that one into its gullet, sensing its life falter as its tissues dissolved. It could feel its unborn young settle through their nascent electromagnetic signals. Their satiation grew. The parent reached for the third victim, thinking maybe after this meal there'd be time to enjoy the last one.

The young were quiet now. It consumed the third too rapidly, but not as rapidly as the first two. It sensed waves of fear and anguish from the fourth. It decided it could take its time.

The kraken reached out for the last one, holding it gently, tearing away the layers. Then, slowly, its tongues slid languidly over the prey's body, rasping the flesh to release delicious flavors. It pulled the prey's feet into its digestive organ. As it slowly consumed this prey, it ceased suppression, augmenting its experience as it looked into the prey's watering eyes. It fed, slowly and gloriously. Then, just as it began to feel the prey's life ebbing, something happened.

The parent kraken suddenly didn't care! It wasn't hungry anymore! As long as it could remember, the kraken felt varying levels of hunger, of desperation, but it had never *not* felt hungry! Instead, it wanted to find a quiet place, a safe place to rest.

Abruptly, the gorged parent pulled the prey into its digestive organ, then began hunting for such a place.

DAY 17
RESPITE

As Yoshi refilled his coffee mug and browsed breakfast options, he heard Jane's voice.

"Make good choices!"

He turned to Jane's smirk, tilted his head side to side, and retorted, "Look who's offering advice!"

She nodded. "Do as I say, not as I do!" She reached to refill her mug as well, then selected a doughy pastry drenched in glaze. She gave a couple of eyebrow raises to Yoshi, who smiled and shook his head, then turned and gestured towards Pavel with her chin.

"He's close to burning out."

Yoshi sighed. "I know, I know. But, he won't hand it off to any of his people. Says we're too close, that it's almost over. So, yes, he *is* stressed, but he's functioning." He shrugged. "He does this. As do you, y'know. Once again, there's irony having such advice come from you."

"Fair enough. Hey, it's part of my job to point out this out to you, and part of your job to decide whether to ignore me.'"

Yoshi pursed his lips. "I'm keeping an eye on him. How's *your* team?"

Jane's shoulders sagged. "Demoralized. There aren't many wounded to treat, and very few bodies to recover. Mostly kraken killed while feeding."

She shook her head. "I know we all looked through Pavel's modeling. I remain convinced that, overall, we've lost fewer people by focusing on killing the things rather than trying to rescue victims, but… She paused, considering her next words carefully, then nodded towards Pavel again. "I suspect a lot of Pavel's squads are going to need some counseling when this is all over, assuming any of us survive!"

She sighed. "We've lost people, Yoshi. A lot of my emergency response people volunteered to go rescue victims, and that put them at risk. Honestly, I don't know if they were any more at risk responding than they would have been in their quarters. How could you tell? But, some of my people went out and were never heard from again. But what gets me? What gets me the most are the people that are missing that we don't know about yet! And, we won't know until this is all over and people notice who's *not* coming out of their quarters..."

Yoshi nodded, his head bowed, his face despondent. There wasn't much to say. He'd heard similar things from the others and had similar thoughts of his own. But, he knew people dealing with trauma needed to express these feelings; there wasn't anything he could say or do that would really help beyond just listening.

He felt confident that they were, finally, making progress. There was light at the end of the proverbial tunnel. But he also knew, once the kraken were dead, the real work would begin. Yoshi sighed and realized he should talk with Elke about this, see what advice she might have to offer.

▪ ▪ ▪

"I think they may be turning the tide!"

Elke, overwhelmingly relieved, fell into her chair. "Tell me more!"

George took his seat as well. "Latest update from Keith. They've been able to shift from reacting to sightings to an orderly campaign. Drones are patrolling their residential dome, and they've been very effective. They've killed another eighty-seven kraken, big ones, that were hunting the open areas and the residential roof top. More than half were mating.

That allowed the squads to focus on the structure itself. It's a big arc like ours, except, of course, theirs is much further along. We have less than a quarter circle, and they have almost a full circle. What they've done, and I think this is brilliant, is split their squads so they can start at the two ends and meet in the middle. The electromagnetic devices make it so much easier to find the kraken. And, any they can't kill, they drive forward. As they reach the middle, they'll be able to destroy a lot of kraken. Any that leave the structure are picked off by drones. Bottom line? They should have their residential dome cleared later today!"

Elke leaned back and looked past George towards the mountains. She remembered how good it felt to believe the worst might be over. Nodding, she turned to meet his gaze.

"What about the rest of the colony?"

"That's the question, isn't it? The primary dome seems clear as well. The crawlers haven't found any in the last twenty-four hours, and there've been no sightings. They patrolled with sniffers and found no sign of them. When they're confident the residential dome is clear, they can get some of the electromagnetic equipment to the primary dome and repeat the search."

George sighed. "Then there's the rest of the colony. They've almost certainly lost a lot of farmers and livestock. They don't know how many of their agricultural domes have been affected, but they do have a number that are not responding to queries. Not a good sign…And, there were some sniffer samplings from some of the infrastructure spaces. So, once the residential and primary are clear, they'll need to organize themselves for the next phase."

George held out his hands. "I reminded them that their situation is very different than ours was. We were limited to our orbital platform, and they have a sprawling surface settlement, some of which has been in service long enough to no longer exactly match the original engineering plans. Those deviations between the plans and the 'as builts' create a lot of risk"

He shrugged. "I certainly don't want to dampen their enthusiasm, but a premature celebration or relaxation could prove disastrous!"

DAY 17

SLUMBER

As the impregnated kraken made its way through the corridor, it felt electromagnetic signatures that, until recently, would have been irresistibly enticing. Not anymore. It had no recollection of ever experiencing such enticement. Instead, such signatures embodied risks that must be avoided.

So, it roamed. And, as it roamed, it reached out with all of its senses. It wanted, it *needed* solitude, safety, indicators that nothing remotely threatening was present or likely to be.

The kraken found itself in an oddly familiar space; it couldn't remember feasting here just a few hours ago. It's only concern, now, was safety and seclusion. It explored the space. It entered one of the smaller, darkened rooms, ignoring the shredded, bloodstained dinosaur pajamas beside the upturned bedding.

It turned, claws facing the only opening, and sheathed its sensory cluster. Now, for the first time since emergence, the kraken went to sleep.

■ ■ ■

Another wary kraken made its way through the corridor, then passed through an opening. It found itself in the open, surrounded by foliage. Such exposure was *not* acceptable!

It pressed forward, into the foliage, leaves passing over its skin; small branches broke off against it, some rubbing uncomfortably and others drawing blood. It pressed forward, sensing the electromagnetic signatures of tiny things, things unlikely to threaten it. It pressed forward until it came to an impenetrable barrier, a vertical surface that extended well above it.

The kraken pressed forward, along that barrier, until it found a quiet place. It settled its body to the ground, relaxed its legs, withdrew its sensory cluster, and, for the first time in its memory, went to sleep.

■ ■ ■

Everywhere, back through the passages and corridors once deemed safe, impregnated kraken scurried and roamed, looking for places to rest. Small creatures, little vermin, the few survivors of the kraken's previous bloodbath, now passed by safely. All the kraken cared about was safety and seclusion.

They roamed the darkened places, found quiet niches, settled down, and they all fell into a deep, undisturbed sleep.

PART 3
HERE BE DRAGONS

DAY 18

AWAKENING

Dreaming...

The dozing dragons had no notion of how long they'd been dreaming, no sense of the passage of time. There must have been a time before dreaming, but they gave no thought to this. As they dreamt, their brains and bodies grew. Their dreams and their movements became richer, more complex, more sophisticated.

As they stirred, and as they dreamt, the dragons became aware they were not alone. They could sense the presence and movements of siblings. And they could sense they were being carried, carried by something similar to themselves, yet profoundly different. All of this infused their dreams, yet none of it interrupted their dreams.

Except hunger.

As the dragons dreamt, their hunger waxed and waned. As their hunger waxed, their movements increased. Sometimes they felt the one carrying them slow and stop. They would sense that one's relief, and sometimes pleasure, and then their hunger waned. Sometimes, as their hunger increased, so too their movements increased. The one that carried them would not slow or stop. They could feel that one's hunger growing, too, and that one's anxiety and impatience. Sometimes, when their hunger intensified, the dragons would sense fear. Fear, because when hunger was no longer tolerable, there'd be a reaping. The strongest would feed on the weakest, and the number of dragons reduced.

The reaping would continue until the hunger waned.

Once the hunger waned, the fear would subside, and the dreaming would resume. Until, one day, it didn't. One day, there was motionlessness. Not just slowing, but a complete stop.

Whatever carried them was no longer moving.

Whatever it was, the sleeping young could no longer sense its presence. Its electromagnetic presence dissipated, then disappeared. And they knew, somehow, that whatever carried them would never carry them again. Their source of sustenance and protection would never move again. And, they knew, also, that their growing hunger would not ease. Not this time, not without action. But they didn't reach out to consume each other.

This time was different.

They reached out with their forelimbs, feeling resistance. They extended their claws, piercing, then tearing the tissues that surrounded them.

They extended the claws of their hindlimbs, piercing more tissue, securing purchase as the claws of their forelimbs sliced and tore. They pulled themselves and pushed themselves, tearing and slicing, with greater and greater urgency, until they no longer felt resistance. They set their claws and pulled themselves out of their womb and into the air.

The emerging dragons felt a slight chill as the moisture on their skins evaporated, leaving a sticky residue. They stretched their limbs and their wings, stretched their tails, and unfolded their sensory cluster. They opened their eyes gradually, becoming accustomed to the light, making sense of what they could see. Their other senses collected scents and sounds and vibrations, and their electromagnetic organs detected each other.

But, after a few moments, as the novelty wore off, their hunger reasserted.

The dragons realized their claws were set in the meat of their parent's body, and their hunger drove them to feed. They tore at the carcass with their claws. They opened their beaks to rasp at the meat with their tongues. They sliced through the flesh with the edges of their beaks.

The dragons feasted, and they grew.

DAY 18

EYE OF THE STORM ...

Pavel stepped back from his console and scanned the room for Yoshi, rubbing his eyes. He needed sleep, badly. He spied Yoshi refilling his mug, made eye contact, and walked towards him. As he approached, Yoshi motioned to an empty cup. "Coffee?"

Pavel shook his head; he winced as his stiff neck subtly cracked. "No, I don't think I can take any more stimulants." He jerked a thumb over his shoulder and added, "We haven't seen a kraken in over an hour. I've told the squads to take a breather, but to stay ready. Kim's people are recharging and rearming the drones."

Yoshi nodded, then raised his eyebrows. "Any luck coming up with ways to isolate domes?"

Pavel held out a hand and wiggled splayed fingers. "Yes and no. Here's the thing…" He tapped his wrist computer to bring up a wireframe of the colonial domes. As they walked over, he continued, "Most of the inter-dome pathways are the subterranean shuttle tunnels, with infrastructure connectivity running either parallel to the tunnels or inside the tunnels. We've been building them for more than thirty years, and even though the plans always call for parallel conduits, some of the engineers, over the years, apparently ran them inside." He smiled and wagged a finger towards Jack, who was working at his console. "One thing I've learned in this job is to check whether the as-builts match the plans!"

He shrugged. "Anyway, most domes have direct shuttle connections to the main transportation hub here, under the base station. The aggie domes…"

Yoshi grimaced and interrupted him. "Please don't call them that! I don't know why, just irritates me."

Pavel bristled briefly, then took a deep breath. He knew that Yoshi tended to be informal overall but preferred proper terminology during official conversations. And he'd been so focused on his responsibilities he hadn't been paying enough attention to Yoshi's stress level. They were tired. They were *all* tired. But, Yoshi was right, they should maintain discipline given the lives at stake.

"Sorry, I knew that. Anyway, the original *agricultural* dome has a direct connection to the primary dome and serves as integration point for the rest of the agricultural domes, so the farmers at all the newer domes have to make two stops. But, a lot of their traffic is farm to farm and they're central tower stores carry most of what they need..."

He shook his head. "All the infrastructure, like I said, runs through or beside these tunnels to the primary, where there's an integration junction and then a main that runs along the shuttle tunnel to the infrastructure dome. So, where does that leave us? There's not much we can do with the conduits, but they're tight. We probably had a lot of little kraken spread through these conduits, but bigger kraken wouldn't be able to get back. So, I think we can ignore them.

As for the shuttle tunnels, well, I talked to Kim and Jack. We could just send shuttles to the midway points and park them there. Within the tunnels, there isn't much room *around* the shuttles! Well, the tunnels with internal conduits have a little more room, but it shouldn't be much. It's just hard to *know* without looking since the as-built doesn't match the plan. But Kim and Jack've seen 'em and they don't think it'll be an issue.

So, we went ahead and moved the shuttles into the tunnels. We should be blocked between primary and residential, between residential and agricultural, between the agricultural domes and their hub, and between primary and infrastructure. Now, here's the downside. If we want to redeploy the drones after Kim rearms them, they'll need to carry them through the tunnels and shuttle cars. Remember, the domes are 2 km wide and there's some space between them. So, without the shuttles running, it's a fuckin' long walk! Same for any squads we redeploy. So, this will slow things down."

Yoshi nodded and patted him on the back. "Excellent work, thank you!" He sighed deeply, then asked, "since you went ahead and did that, I'm thinking you don't think we're done."

Pavel tilted his head side to side before responding. "I'm *hoping* we got them all. But it's hard to be sure. The sniffers are still picking them up, but that doesn't mean anything. Their scent would linger for a while even if they are all gone. But, we're also not picking up their electromagnetic signatures, and that's what's got me hopeful."

"Did you see a decline? Does it make sense that we got 'em all?"

"Yes," Pavel nodded, "encounters have been declining, but it still feels abrupt."

Yoshi hesitated, then leaned in and looked Pavel in the eyes. "What's your gut telling you?"

Pavel snorted, looked away, shook his head, then met Yoshi's eyes again. "That it ain't over. That another shoe's gonna drop."

Yoshi nodded, then patted Pavel and the shoulder. "Yeah. I have the same feeling. Let's get everybody together."

Yoshi looked around the table. Jane yawned and rubbed her eyes, her power nap prematurely interrupted. Jack stumbled over from his cot, Kim from the coffee machine. Yoshi realized Keith was missing and asked Jane where he was. She paused before answering; maybe they weren't being as discrete as she thought. She shrugged as nonchalantly as she could and her flushed cheeks behind her steaming coffee mug. She looked down at her wrist, tapped her computer, spoke into it, and mumbled, "on his way."

"Where was he?"

"Labs, down the hall, not…" She was interrupted by Keith as he entered the room and made his apologies.

Yoshi nodded, then said, "Pavel tells me we haven't seen a kraken in over an hour. We have, what, over 200 confirmed kills?"

Pavel checked his wrist computer. "236 so far."

"Right. 236 confirmed kraken kills. That's great, but we can't know how many were out there, nor how many were lost to other factors like cannibalism. But, our preliminary estimates were higher than that, right? On the other hand, Pavel, you said you haven't seen a kraken for over an hour?"

Pavel checked his wrist computer. "It's been seventy-seven minutes. And, before that, a definite decline in the rate at which we were finding and killing them, which supports the idea that we might have gotten them all…"

"But you don't think so?" Kim interjected.

Pavel looked over at her. "I don't think so. Doesn't feel right. I can't explain it, I can't defend it, but it doesn't feel right."

The team was silent for a moment, then Jack volunteered. "We are still picking up their scent, but it'll be days before that's fully dissipated. And, we're still picking up their sex pheromones, but that's also dissipating. Here's what's interesting. We're not picking up their electromagnetic signatures."

"Anywhere?" Yoshi asked.

Jack sighed. "We're not picking them up anywhere we're looking, but we're only looking under the primary and residential domes. We just don't have the equipment or personnel to look anywhere else. But, if we're not picking them up where we're looking, maybe we can afford to reallocate some of that equipment and personnel to other parts of the settlement."

"It may be time to add those spore filters to the life support system," Kim added. "And I'm thinking we should reclaim some of our fabrication capacity to make spore scrubbers for the drones. I just wish there were a way to put up spore filtration barriers between domes…"

Yoshi turned to Keith. "Any expert guidance?"

Keith blushed and ran his fingers through his hair. "I uh, I was afraid you were going to ask me that! Of course, I've been exchanging communiques with our Eden friends. They offered a few things we should think about. One, our kraken emerged from a number of hosts over a series of days. They mature as they grow, and they grow as they feed, so this may not be a factor, but we should bear it in mind. Second, kraken don't live very long. They emerge, they feed and grow rapidly, and, after a few short, awful days, they mate and go dormant until their young emerge. Finally, when they go dormant, their electromagnetic signals shut down. So, there are two possible reasons why you're not detecting them. One is that they're dead. The other, more likely scenario, is that they're dormant and their young are about to emerge."

Yoshi nodded. He'd experienced a flash of optimism he knew was unrealistic. Sure, it was theoretically possible they'd killed all the kraken, but it was extremely unlikely. Inwardly, he chalked it up to exhaustion. He was desperately tired, and knew his team was, too.

Tired people make mistakes, and mistakes cost lives.

Well, his people would take their cues from him. He couldn't do much about their exhaustion. But, he could try and model the commitment and discipline it would take for all of them to prevail!

"Alright, people!" Yoshi exclaimed. "I'd love to be proven wrong but it sounds like this isn't over! Keith, if we have dormant kraken, what can we expect next?"

Keith glanced nervously as all eyes turned towards him. "Okay, well, it's probably better I repeat things you know than leave anything out. But, I can summarize if need be!"

Yoshi raised a hand to interrupt him and made a sweeping gesture towards the congregated staff. "I think I can speak for everyone when I say we're all exhausted. And, the last time you walked us through this, we were a lot more focused on the kraken generation. So, yes, better to repeat than omit. Go on."

Keith nodded and took a deep breath, like he was trying to remember every little detail. "Eden's science team has been studying kraken since their outbreak fifteen years ago, and they've learned a thing or two that'll help us. Remember the life cycle: spores, parasitic worms, kraken, then dragons that produce more spores. The spores, worms, and kraken are haploid, with one copy of each gene. When kraken mate, they merge their genomes with those collected from their mates to produce diploid dragons that, in turn, produce haploid spores.

Remember, also, that the kraken are succession pioneers. Earth had a number of species that specialized in exploiting the aftermath of disasters, fires, floods, quakes, whatever. But kraken have them all beat. All the kraken we've seen are the descendants of one spore that spread through one man's body as worms and then were passed on through intercourse! Which would mean that all the kraken were genetically identical, but the way they mix and match genetic code during fertilization still creates genetic diversity in the dragon generation, plus the manipulation of genetic expression in the womb…"

He scanned everyone's faces and realized he'd lost them. "Oh, uh, sorry, I'm tired, too. And, though I was aware of Eden's life forms before all this started, I've learned so much more trying to help us prepare. Yes, these things are absolutely awful, but they're also amazingly adaptable. I had no idea!"

Yoshi met his eyes. "We know you're tired. We're all tired. And, yes, if this were a biology class, we'd all find them fascinating. But, we really need you to focus on what can help us."

Keith blushed. "Fair enough, and my apologies." He cleared his throat before continuing. "Right now, assuming, for our safety, we did not kill all of the kraken, some number of gravid kraken (and we can't know now many) have sequestered themselves and shut down. The parts of their nervous systems that maintain the body, breathing, pumping blood, that's still running. Everything else has shut down, permanently. They do this when their young are ready to emerge. The young aren't born, there's no pathway out of their parent's womb. Once the kraken are dormant, the young *eat* their way out, then consume their parent's remains. We can expect that, right now, a number of dormant kraken have produced offspring, and those offspring will, once they've finished eating their parents, start hunting."

He looked around the room.

"As we've discussed, the dragon offspring are highly variable in terms of size and the level of cooperation between siblings. We're likely to see loners, and we're likely to see highly interdependent packs. Some packs will just cooperate with each other, and some will be outright altruistic, especially when they derive from the same zygote, uh, fertilized egg, because they're genetically identical. They'll be willing to give their lives to help their pack mates release spores. That's a level of altruism we see on Earth among ants and bees and termites. And, like termites, some of the packs may even have castes, specialists, elites, soldiers, pawns, with some degree of physical specialization. There's just no way to know!"

He brought up a projection of a dragon. "The dragon generation shares some characteristics with the kraken generation, but also has some profound differences. The body is leaner. The beak is stronger and holds nine tongues instead of three. The forelimbs end in talons instead of claws. The midlimbs are leathery wings, and they are adept flyers. They can rise vertically, hover, and fly backward, at least until they get very big, at which point the wings are used for balance. The hindlimbs are stout and powerful, and when the dragon isn't flying, they walk on their hind legs with wings and tails providing balance. Or, they go on all fours, walking on the knuckles of their forepaw claws.

They have the same sensory capabilities as adult kraken, including their electromagnetic capabilities. They have comparable camouflage capabilities to adult kraken. They can match colors and textures, and they can dampen neural functions to make their camouflage more effective. Eden's research team believes the larger, solo dragons use their electromagnetic capabilities like the kraken generation. The smaller, pack hunting dragons don't seem to bother unless they have to. They just physically attack and overwhelm their prey.

They're smart, smart as a dog or octopus. They can solve problems. They are, if anything, more aggressive hunters and feeders than kraken. Kraken like to play with their food, doing everything they can to keep their prey alive and take pleasure from maximizing their suffering. The dragons eat quickly so they can hunt for more food. They never sleep, never rest, never stop hunting. The kraken were in a race to reach breeding size, and the dragons are in a race to produce as many spores as possible. At some point, they'll have millions of spores ready to go. As that time approaches, the dragons will slow down and seek a place to hide, just like their kraken parents did. Once they release their spores, they'll die, too."

"How big do they get?" Kim asked.

Keith shrugged. “No hard rules on that. Some may start out as small as a cat, others considerably bigger than a human. The bigger they get, the more spores they can release, so they’ll grow and defer sporulation as long as they can. But, Eden lab experiments show that, if they feel sufficiently threatened, including picking up alarm pheromones. or experience a debilitating injury, they’ll release whatever spores they can, then die. The altruistic packs may even communicate electromagnetically, thereby triggering the whole pack to sporulate at once.”

Pavel whistled. “So, once we start killing them, that’s likely to cause the others to release their spores?”

Keith nodded. “Death or significant injury results in release of alarm pheromones, and that affects other dragons.”

Kim started tapping at her wrist computer and asked, “Do you have the alarm pheromone? If you do, we might be able to filter it out?”

Keith nodded and said he’d get it to her, adding, “This needs to be a top priority. We have to kill these things, and, as we kill them, they’ll release alarm pheromones.”

“Would the spores infect surviving colonists?” Jane asked.

Keith nodded wearily. “Yes, any spore could infect a colonist, a dog, or cat, anything really, and start the whole nightmare over again. And the spores can breach many of the polymers we use. That’s how the first Eden colonist was infected; a spore penetrated his environmental suit. Well, we use the same polymers in many of our infrastructure components. Now, we discussed this a while back. Such breaches would be pin pricks, and we’re not dealing with a significant pressure differential, so we’re talking about a maintenance headache but that’s about it. The issue here is infected hosts.”

Pavel leaned forward. “How do we find the dormant kraken?”

Keith shrugged. “You probably don’t, at least not in time. They could be anywhere under a dome, anywhere in the infrastructure beneath or between domes, any dark, quiet place a kraken could shove itself into. Sniffers could help. Motion sensors? No. Electromagnetic sensors? No. And you don’t have much time, right? The dragons will emerge very quickly. Still, any you can find means that many fewer dragons…”

Jack asked, “What does this mean for reassigning squads from the residential dome?”

Pavel tilted his head side to side, then said, “I think we wait. I think we wait until we see where these things are emerging. We can’t afford to have these things rampaging through the residential dome." He turned back to Keith and asked, "weapons? Still flamethrowers? What about the drones and electromagnetic detectors and disruptors?"

Keith paused, then said, "Certainly the flamethrowers and drones should work as well against the dragons as they did against the kraken. The electromagnetic devices should detect the dragons. I'll need to ask the Eden folks if they were able to drive dragons away. Makes sense that it would, but I have no idea whether the frequencies that worked on the kraken will work on the dragons. I'll check!"

Pavel sighed. Turning to Yoshi, he said, "I think we're in the same situation we were in with the kraken. I think we have to prioritize killing dragons over rescuing colonists. We need to do everything we can to prevent these things from releasing spores."

Yoshi silently polled the others, meeting their eyes one by one. He turned to Pavel and said, softly, "whatever it takes…"

As Pavel walked away, Yoshi turned to Kim and Jack. "We're going to do whatever it takes. We're going to ensure Pavel and his people have whatever they need. But, my confidence that we'll manage to kill every dragon in time is very low."

They nodded, so he continued. "Here's what I need. I need a spore cleanup plan. I know you've added spore filtration to life support. That will help, but we still have spores moving through the domes and settling on surfaces. Let's say spores are released under a dome. How do we reclaim that space?"

"We're building scrubbers for the drones," Kim said. "It means taking those drones from the dragon killing contingent, but they can filter spores out of the air. The spec Eden sent also includes having these scrubber drones going over some surfaces, like the insides of our domes. But, at some point, we're going to need a way to clean interior surfaces."

"I need you two to partner," Yoshi pointed back and forth between Kim and Jack. "I need you to assign some of your best people to go huddle somewhere and get me an answer. Okay?"

DAY 19

EMERGENCE

As the parent's body was reduced to a skeleton, the alpha reared up and extended its sensory cluster.

The dragon spawn huddled in a dimly lit, enclosed space. It felt no vibrations; it heard no noises nor picked up any scents, other than those of its feasting siblings. It knew each sibling by its electromagnetic signature. It had no concept of numbers, but, if any of these ten signatures ceased, it would know. It reached out with its electromagnetic senses. The enclosure consisted of flat surfaces, and one of them had an opening.

The alpha considered its many siblings and decided their least bad option was to risk sending one through the opening.

Its sibling, frustrated to be dispatched before their parent's carcass was completely consumed, acquiesced. It spread its wings, and flew towards the opening. As it approached, it changed the pattern of its wing movements to hover, then slowly moved through the opening. The space beyond had more light, allowing the sibling to see that it too was composed of flat surfaces. There were a number of objects on the ground, including fecal matter from their parent and strips of material that bore the fading scent of prey. Continuing to hover, it moved about the room until it was satisfied that the room was safe but held no sustenance.

The alpha, seeing the parent's body offered no more nourishment, stretched its wings and flew through the portal, following the one sibling and followed by the rest.

■ ■ ■

Yoshi stood by his window, steaming cup in hand, and let his mind wander. He could hear the murmurs of his team as they worked. He knew that, with a bit of effort, he could follow their conversations, but he didn't. They knew their jobs. They'd let him know if he was needed. For now, the murmuring was a soothing background noise.

Their apparent respite was welcome, but he knew it was temporary. He'd read Elke Lubandi's recounting of Eden's kraken ordeal. He couldn't imagine facing these creatures with no idea what to expect. He never fully appreciated what they went through, all within a few weeks of their arrival! Alexandria, at least, had the benefit of their experience and their guidance through this crisis. On the other hand, Eden Colony was so small, contained on their orbiting platform. Their problem was, for good or ill, bounded.

Alexandria was huge. So many places a dormant kraken could be hiding. They couldn't possibly know how many of these dragons they might be facing. So yes, they had the benefit of Eden's experience to guide them, and they were doing, he believed, he hoped, everything humanly possible. But, how could they *possibly* be adequately prepared?

■ ■ ■

The dragons were increasingly agitated.

Their parent was picked nearly clean, and they all knew, once the last morsels were gone, larger dragons might turn on their smaller siblings. Prudence suggested the smaller ones forgo the relatively easy feeding of their parent's remains and leave sooner rather than later.

They were in a darkened room. Objects they did not recognize, a bed and a dresser, had been shoved aside by their parent as it sequestered itself in this room post feasting in the residence's common area. The dragons neither knew nor cared about any of that and took no notice of the bloody strips of clothing as they made their way, one at a time, through to the corridor.

Some went left; others went right. They moved along, continuing to make choices: left, right, up stairs, down stairs. Famished and keen, they put distance between themselves and their largest siblings. The migrating dragons sought the electromagnetic signatures of anything large enough to satisfy their growing hunger.

One dragon stopped. It could feel, in the distance, a faint signal. Slowly, it moved forward. There were no other siblings nearby, which was good. There'd be no competition. As it moved forward, the signal strengthened. There were two, no, three creatures nearby. It came to a corridor junction. It moved forward, then hesitated. It backed up, then moved to its left, then to the right. The signal got stronger and stronger.

Then it got weaker.

The dragon stopped, turned, and realized the signal was coming from behind a barrier. It pressed the talons of its forelegs against the barrier, raising its beak and sensory cluster. The barrier felt firm and unyielding. It remained motionless, feeling the signals coming from beyond the barrier. It experimented, sending signals of its own, sensing how the creatures on the far side of the barrier responded.

It learned, slowly but surely. It refined its signals, gently subduing the creatures on the far side of the barrier. They were no longer moving. This was good, but pointless unless it could find a way in.

The frustrated dragon paced. It was hungry. It wanted in. Suddenly it stopped, for it noticed something. There was another barrier above, a different kind of barrier. It raised one of its forelegs, extended a talon, and pushed against the overhead barrier. Part of the barrier moved. It pushed again, and a whole section of the barrier shifted. It pushed that section aside.

The dragon extended its sensory cluster such that the ends of its eyestalks could scan the darkness above the barrier. Light glowed through the horizontal barrier on the far side of the vertical barrier separating it from the tantalizing electromagnetic signals. It pushed more of the panels aside, then extended its wings and began the complex beat that would lift it vertically into the darkness. It moved forward, hovered, considering the panels and the light glowing through them.

The dragon extended its forelimbs and hindlimbs, tucked its wings, and fell through the panels to the room below. The panels, some broken, scattered across the floor. The dragon reared up, its eyestalks scanning the room.

There were three creatures in the room. As it fell through the panels, it could see that they were vertically oriented, then scattered to get away as the dragon fell. Now they stood silently. It could feel their fear.

The dragon decided.

The dragon reached out with its talons and grasped one of them, the tips of its claws piercing the prey's flesh. The others stared as it lifted the one it held. It ceased subduing that one, and the prey began to thrash and make noises that the dragon found stimulating. It focused four of its eyes on the one while scanning the others with its two remaining eyestalks. It opened its beak and reached out with nine long tongues, exploring the thing it held, probing and tasting.

The thrashing prey was enclosed in some tasteless material that tore away easily, exposing tasty flesh. It probed some more, pressing its tongues against the prey's skin. The flavors were absolutely delightful. It lifted the prey higher, grasped the prey by the head with its tongues, and pulled it into its beak. The prey squirmed and screamed as the tongues pulled it into the dragon, its beak closing around as it felt the prey press into its digestive organ. As the juices flowed, as the prey stilled, flavors exploded.

The dragon turned, feeling intense fear from the one on its left. It reached out for that prey, lifted it, stripped it, and swallowed it also. It turned to the last one. It eased suppression. The waves of fear coming from it grew and grew, and it started to move away. The dragon found this stimulating and moved towards the prey as it backed away. The prey backed up against a vertical barrier; sliding sideways, it shook its head side to side, emitting delightful noises. The dragon enjoyed this, but it was hungry, and this was not as important as the release. It abruptly shot out its forelimbs and crushed the prey's body as the dragon lifted it. The prey hardly resisted as the dragon stripped and swallowed it.

The dragon scanned the room once more, then reached out with all its senses. It felt another electromagnetic signature in the distance. Extending its wings, it lifted itself through the hole in the ceiling and moved towards the signal.

■ ■ ■

When their parent's body was picked clean, the alpha spread its wings and raised itself, reaching out with its senses. It felt the electromagnetic signatures of its siblings as they jostled for the last tidbits. It had no concept of numbers, but it knew the unique signature of each sibling. There were two elites, fourteen soldiers, and twenty-seven surviving pawns.

The alpha considered their options, then dispatched four pawns through the opening. It could feel, through their electromagnetic signals, that neither prey nor predator waited for them in the next space. The alpha flew through the opening and hovered, reaching out with its senses, sending pawns in each direction to extend its sensory range.

There! Prey nearby!

The pack flew as a cloud through the corridor until they reached the signal. They hovered by a barrier. Some settled to the floor. All used their forelimbs to poke and prod the barrier, seeking a way through. One of the soldiers that settled to the floor focused on an electromagnetic field covering an opening in the barrier. It probed and prodded electronically, until the field flickered, then failed.

■ ■ ■

Henry sat at the kitchen counter talking with Omar. They'd seen the recommendation that smaller families shelter together. Something about limits to the number of individuals the kraken could handle at a time. So Henry reached out to some friends, who decided Henry's place had the most room, and now all the kids are bouncing on the living room furniture as Henry, Omar, and Sid speculated how much longer they were going to have to do this. Because, friends or not, they were starting to get on each other's nerves. Sid's wife sat by herself, refusing to talk to Omar and Henry's wives, who told Sid he could bring his damn dog anyway! It was one of those little fluffy things that constantly barked its shrill little bark. Henry swore he was getting ready to strangle it.

Henry stopped talking. Something changed, but he wasn't sure what. He realized everyone had stopped talking. The kids had stopped bouncing.

He looked around, then noticed that the portal was open.

How had he not noticed it sooner? The locked portal made a very light humming noise, nothing one would pick up from the bedrooms, but something subtly reassuring to hear before bed. That humming had ceased, and the portal was open, and he started to get up to go close it again when all of a sudden *things started flying through the portal!*

Everyone froze, then screamed, then ran, batting away the horrid things that flew through the portal. They were small, about the size of Sid's dog, but there were so many of them! Henry saw two of them fly into one of the smaller kids, knocking her off her feet. As she fell, the things dug their talons into her, opened their beaks, reached out with their tongues, and pulled part of her into their beak as they bit down and sliced through her flesh.

The kids were screaming, and their wives were trying to drive the things away, but the swarm knocked the women down to eat them alive as they tried to fight back. Still frozen, Henry saw Sid and Omar run past him, Omar with a kitchen knife and Sid with a spatula (which was so stupid of Sid, but at least he was moving).

Henry looked around, saw a frying pan, grabbed it, and ran into the living room. He swung that pan with everything he had, batting the little nightmares away from one of the kids; but, the kid was a bloody mess and not moving. The kid's blood pooled instead of pulsed. Some part of his brain told him that meant the kid was gone, so he started swinging at another one attacking Omar's wife, who was still trying to fight back.

Something stabbed him in the shoulder, and the impact made him stumble to the left. But, then something else stabbed him in the thigh and another thing took a massive bite out of his ass. He fell to the ground. He was face down, trying to get his hands under himself, looking for the frying pan that had gotten away from him.

Then Henry saw six eyes look into his own; he felt knives dig into the sides of his head. He saw the beak open and several tongues reach out. The tips of the tongues moved smoothly across his face to the sides of his head and neck, then pressed against him. A thousand needles dug into his skin. He started to scream, and the beak opened; the tongues pulled him into the beak and he felt the sharp edges of the beak cutting into the sides of his head and then…

DAY 19

REPORTS

Amy wasn't any less exhausted than the others, but she couldn't sleep. She'd tried! She laid down in the shade on the grass and tried to close her eyes, to no avail. It wasn't just that some part of her expected to be shooed off the grass. Tired as she was, she couldn't turn off her mind. Or, at least, not the part of her mind replaying the awful things they'd seen the last couple of days.

The grass surrounded a deserted café. She'd been a barista, briefly, a lifetime ago, and still knew her way around an espresso machine. A few minutes later, she walked past her squad to a picnic table, settled in to enjoy her latte, and let her mind wander. Well, semi-wander. Here, as opposed to drifting off to sleep, she had some control over her thoughts.

It was a lovely afternoon. Stirring her latte, Amy surveyed the area. The café was on one of the main boulevards connecting the central tower's transportation hub to the main entrances of the residential structure. From here, she could see the construction project that would complete the circle. She pondered how many additional residences it would provide compared to their immigration rate, and when their second surface settlement's residential dome would be inhabitable. Then, she realized the other variable that would have to be factored into the equation was this incident's casualty rate.

She shuddered, shoving those thoughts aside.

Instead, she pondered the shadows slowly stretching before her. It felt "right," though she had no personal experience to measure against. Amy used the PIP program to get to Alexandria for a free college degree. Her birthworld was a lifeless rock with a pressurized dome. At least Alexandria had an atmosphere, such as it was. But, she'd never experienced natural lighting. Artificial supplementary lighting under domes, artificial light aboard a deep space vessel, but never natural lighting.

Still, the lighting felt right. Shadows ever so gradually lengthened as she sat in the cooling shade of a maple. A gentle breeze ruffled its primarily red leaves, though some remained green. She'd never given the breeze much thought, but she presumed that too was artificial, all part of creating the illusion of what her ancestors once took for granted on Earth.

The things she'd seen…

Amy shook her head again. She sighed, then idly stirred the dregs of her drink. Deciding whether to make another, she suddenly saw something above the residential structure to her left. She turned, squinted, then stood.

Her first thought was flying kraken. Her second thought was they were too lean. Then, she remembered the briefing about the dragons.

There were three of them, coming this way.

Amy stared, transfixed, then dropped her mostly empty cup onto the grass and called out to wake her squad.

▪▪▪

"I've got a sighting!"

Pavel reached for one of the projections above his console and dragged it to the conference room table. As the others gathered, Pavel expanded it. They could see the café and surrounding area; at the edge, a trio of dragons flying towards the café.

Yoshi pointed a shaky finger at them. "How big are they?"

"Maybe 60 or 70 kilos."

"What are they doing?"

"Right now, hovering. But there's a squad right *there*, at that café. Expecting the dragons to notice them at some…"

As if on cue, the dragons moved towards the café, accelerating and descending.

"They've seen the squad!" Pavel turned to Kim. "How many drones are ready to fly??"

"Ah, most *should* be, let me focus on getting the ready ones launched, ok?"

Without waiting for a response, she ran back to her console.

▪▪▪

Chad shook himself awake, stood, and started yelling at the others, kicking a few to get their attention.

"Get up! NOW!!"

As each squad member came to and stood up, they also turned to see what Chad and Amy were pointing at. Once they saw the incoming dragons, they pulled on their flamethrowers. The dragons flew low over shrubs and between trees and structures. One lighted about 3 meters in front of them, the second circled and lighted behind them, and the third hovered a couple of meters above them.

Amy shouted, "Form a circle! Disrupters inside!"

The two carrying electromagnetic disruption equipment stood back-to-back. Armed squadmates arranged themselves in a circle around them. As the squad arranged themselves, Chad pointed and shouted, "Anil, you keep your eyes on the one above us!"

Amy called out, "Hold your fire until they're within 2 meters!"

The one dragon hovered above as the other two turned to their left, walking clockwise around the squad.

"I'm feeling a little dopey!"

"Shake it off; they can't suppress all of us!"

"I'm trying to find the right frequency…"

Suddenly, one of the dragons charged. Raising its tails and spreading its wings, the dragon reared up, forelimbs and talons extended. Quickly, it closed the distance and slashed one of the squad members from right shoulder to left hip. As she screamed and fell, the dragon opened its beak, reaching for her with tongues and claws. The squad members to her right and left overcame their shock and doused the monster with their flamethrowers.

The dragon staggered backwards, writhing in pain, then stopped moving.

The dragon pacing behind them spread its wings, lifted itself into the air, and followed the hovering dragon towards the residential structure. Amy walked to their fallen squad member, felt for a pulse, and found none. She hung her head and swore.

■ ■ ■

Pavel hung his head as well. They'd lost a lot of colonists, but they'd managed to kill over 200 kraken without a squad casualty. Until now, kraken hunted individually and were no match for an adequately armed squad with too many members to suppress.

Chad asked what they should do with their fallen squadmate. Pavel paused, looked up, then responded. "What you *should* do is carry her body back here, but we can't do that, not right now. Those things are attacking colonists. I need you…" Pavel paused and swallowed hard, regretting his words even as he spoke them. "I need you to leave her behind and head to the residential structure."

Chad paused, then acknowledged. As the command staff watched the display, the squad, one by one, touched their fallen friend's shoulder and walked along the boulevard.

Jane whispered, "Should *we* go bring her home?"

"I can send a team with disruptors..." Kim said through tears.

Yoshi let Pavel answer. "No, not until we find the right frequency."

"The disruptors should work!" Keith offered. "But kraken operate within a relatively narrow band, narrower than dragons. So, it's not a matter of finding the right frequency, it's more about expanding range during an encounter..." He trailed off as he stared at his wrist computer. As he did, he slowly shook his head side to side, then began to smile.

Mildly annoyed, Yoshi grimaced. "Something you want to share?"

Keith nodded, then looked up. "It's from ..." He paused, beaming. "It's one of my grad students, Nona. She just notified me. She was reading some of the latest reports from Eden and had an idea. She reached out to them, the Eden team, and they've been exchanging messages. They think it'll work!"

Yoshi raised his eyebrows, then prodded. "Enlighten us, please..."

Keith collected himself, then exclaimed. "Self-organizing drones!"

Kim gave him a quizzical look. "Like we use for dome construction?"

In a geologically stable world like Alexandria, the best human engineering could build was a 2 km wide dome. Dome size was a function of drone precision. Domes consist of a thick mucilaginous concrete disk with a raised edge, or stem wall. At the center of the disk was a tower. Its base held the dome's transportation hub, providing access to a network of subterranean shuttle tunnels running between domes. The top of the tower provides the central support for the dome's supporting arches.

Swarms of self-organizing drones lifted arch segments and fit them together. When all the segments were ready, spanning from the central tower to the stem wall, more drones lifted dome sections and placed them on the arch segments, their weight locked the arch sections in place. Feasible dome sizes depended on advances in arch materials, providing greater strength or lesser weight, and in the precision of the self-organizing drones. Everyone in the staff room was familiar with collective drone capabilities, so Keith's student's idea did not seem that original.

"Yes and no." Keith shook his head, then continued. "Yes, in part, they're like construction drones. But they're also like another class of drones the Eden fabrication team developed for their Xenobiologists. They call them 'ticks'. They're self-organizing, like the construction drones, but smaller. They're loaded into special shotgun shells field teams fire at Eden's larger life forms. They make contact, spread over the life form's body, and collect data. Including taking minute tissue samples, which is where they got their nickname. It's really brilliant!"

Yoshi struggled to contain his frustration. "Keith, I'm not following you here!"

Keith raised a hand and apologized. "Nona asked the Eden xenobiologists to ask their engineering team whether the ticks could be modified to find and kill dragons. Here's what they came up with."

He flicked his wrist computer, and a drone appeared over the table. "This is a small drone, only 3 cm side, smaller than your thumb. That will allow it to fly through tight places, which will be important with smaller dragons. They're self-organizing, like our construction drones. They'll fly around the domes and search for dragons. They can sense dragon electromagnetic signatures, home in on them, and decide how many drones they'll need to destroy it. Instead of the little tissue sampler used by the Eden xenobiology teams, they have small blades. A swarm of them will, essentially, shred a dragon. They'll tear up their wings, slice off their sensory clusters, cut through their hides, shred their organs. The dragons can't fight back; the drones are too small. It's just brilliant!"

Jack nodded along. "Yes…Yes, I can see how this would work! But, it'll take my team a while, maybe days, to come up with designs…"

Keith interrupted him, his voice rising with excitement. "No! The Eden team *already has the designs*! They thought of this a couple of days ago, and they're adapting their tick designs. They've even built a prototype! They're sending us the specifications now!"

Jack turned to Kim and asked, "Any idea how long they'll take to build?"

She shook her head, wiping away tears. "Not a clue, not until we see the specs. I'm sure some of the components will be off the shelf stuff, but I'm also sure there'll be unique components. I mean, Eden can send us the specs, but we'll need to set up fabricators to make them. Then, we have to assemble and program. But, I'll let you know as soon as I know!"

Keith added, "Eden team says part of the self-organizing capability is picking their battles. Deploy them as you make them; as soon as there's enough of them, they'll start hunting dragons. They'll start killing smaller ones, then work their way up as we deploy more of them!"

Pavel stood, frowning, arms folded, rubbing his chin with thumb and forefinger. He pointed that forefinger at Keith and asked, "What would happen if we deployed just one?"

Keith looked at him quizzically. "I suppose it would fly about, looking for dragons, but not attempting to engage. Why?"

"Build 'em and deploy 'em. But, as soon as you have enough to start hunting dragons, I want to redeploy a few. Send one each to the aggie domes that have stopped responding. Find out what the hell's going on over there!"

DAY 20

DRAGONS ATTACK . . .

The dragon was hungry.

Walking through the corridors was uncomfortable, but flying outside was dangerous and unproductive. It sensed very little prey, and the prey it did find fought back. Flying with its two siblings, they'd detected the electromagnetic fields of prey. They descended to attack. They were larger than the prey and thought the feeding would be easy, but the prey formed a circle and sprayed fire outwards!

One of its siblings died in agony as it burned!

The dragon, frustrated, furious, and terrified, flew upward with its surviving sibling, leaving a trail of alarm pheromones. Something with a very strange electromagnetic field rammed its sibling; the sibling immediately imploded and fell, its electromagnetic signal fading rapidly, then ceasing altogether. It sensed another strange electromagnetic field heading its way. It folded its wings and dove to the ground, spreading its wings at the last moment. Then, folding them once again, it darted into an opening. The space on the other side of the opening was tight, but the strange thing hovered outside. So, the dragon turned and walked on all fours through the strange narrow canyon.

As it walked, it sensed the electromagnetic signatures of prey. It experienced some anxiety; these prey were like the ones that burned its sibling. Would they, too, be able to fight back? But it was also hungry, needing food to grow and prepare for the release. The release could not be entrusted to a sibling! It *must* achieve the release for all of them!

The dragon followed the signal and came to a vertical surface. It bombarded the prey on the other side of the barrier heavily. When it was sure of their stupor, sure that they could not fight back, it explored the barrier, looking for a way over or around it. Finally, driven by hunger and frustration, it pressed its talons against the surface.

The surface tore! In a frenzy, the dragon raked the barrier with its claws, tearing away the material, exposing structures that gave more resistance but were no match for its strength and desperation. It tore and it pulled and pushed and bent and forced its way through. The space beyond held a number of prey. It could sense their anxiety.

The dragon reached for the first, grasping the prey with its talons, tearing away the tasteless material encasing it with its tongues. Then the dragon lifted the prey and grabbed it with its tongues; it tasted the flesh, and pulled the prey into its beak. The prey wriggled briefly as it was swallowed, making delightful sounds of pain as it struggled. But, the noises ceased as its digestive juices did their work.

The dragon turned, aimed its eyestalks at the next victim, and saw its optical sensors staring back. It saw the prey shake. It eased that one's suppression and saw it shake more. The dragon heard it begin to make shrill noises that were just *so* exciting. The dragon reached for the prey and fed again.

▪ ▪ ▪

Jack approached Kim as she refilled her mug and inspected the snack options. Unintentionally, he startled her as he touched her shoulder.

"I'm so sorry! Did not mean to scare you!"

Embarrassed, she assured him she was alright. "I'm preoccupied. Tired and preoccupied. We should have the first tranche of the new self-organizing drones ready for deployment soon. We had a lot of components in stock. We prioritized fabricating the parts we didn't have, and we'll be restocking the existing parts we used so we can keep assembling them. Programming went smoothly too..."

She faltered as she realized he was waiting to tell her something, then raised her eyebrows and held out her hand to encourage him.

He smiled. "I was working with one of my engineers, and she had an idea. The drones we've deployed, the ones with the flechettes, are not self-organizing. But they can receive instructions! She wrote some code you could add to the self-organizing drones that would let them include the larger flechette drones in their planning. See? When they're going after a larger dragon, they could call in some of the flechette drones..."

Her face brightened and she slapped his shoulder. "That's brilliant! Why didn't we think of that? Get me the code and we'll add it. We have a batch going out now, but if we add your new code to the ones we're programming now, they should be able to distribute that code to the ones we're deploying. Brilliant!"

She picked up her mug, grabbed a nutrition bar, and strode back to her console, muttering "brilliant!"

▪ ▪ ▪

They walked through the tunnel connecting the primary and residential domes. The tunnel's maintenance lights were on, thank goodness. Foot traffic here was rare. Once the tunnel was completed and shuttle operations started almost thirty years before, there'd been no need. It was, essentially, a 3 km long, 8 meter wide tube, just barely wide enough to accommodate the shuttle car that rocketed back and forth between the domes.

They walked along the flattened floor of the tube on either side of the monorail the shuttle hovered above. Two teams of eight guided suspensor pallets filled with deactivated self-organizing drones. Each pallet also carried an electromagnetic disrupter tuned to the frequencies Eden's specialists thought would drive away dragons. They really, really hoped they wouldn't need to test them.

There was, of course, some debris along the way. Vermin remains and parts that fell from the shuttle were subsequently replaced at either station, nothing particularly notable. After what seemed like forever, they came to the shuttle car now blocking the tunnel. Hesitantly, two opened the shuttle car service door, walked in, turned on the lights, and prepared to open the far end service door. The rest of the team activated drones.

As they did so, the drones rose; then, after a brief hesitation, they flew into the car. The two in the car cautiously opened the far door, but there were no signs of dragons in the tunnel. They watched the drones fly through, heading along the tunnel towards the residential dome. When the last passed through, they closed the door. As they exited, they closed the other door and started walking the suspensor drones back to the primary dome hub for the next load.

▪ ▪ ▪

The alpha searched the chambers. Having detected the electromagnetic signatures of numerous prey, it led its siblings through a corridor to a vertical surface. There was prey, lots of prey, on the other side of this barrier.

The dragons poked and prodded for some time before one of the siblings discovered the barrier above them consisted of panels that could be pushed aside, revealing a darkened space that extended beyond the vertical barrier. The alpha, deciding a sibling could be risked, sent one into the darkened space. From there, it could see there were more panels above the prey. Tentatively, it lifted the edge of one as quietly as it could so as not to startle the prey.

The panel moved. With that, the alpha flew up into the darkened space followed by the rest of its siblings. They hovered above the panels; then, as one, the dragons shoved the panels aside and landed within the largest chamber. They didn't even bother suppressing the prey! The prey, reeking of delicious fear, made wonderfully shrill noises as they rushed about the chamber, knocked things over, and bumped into each other.

The alpha saw a larger one, a creature almost the same size as the alpha. Extending its wings and tails for balance, the alpha reached out and grabbed it. The alpha barely had the strength to lift the prey as it struggled. It used its tongues to tear away the layers of inanimate material; then, slowly, as the prey continued to struggle, blood seeping from where its talons pierced its skin, the dragon ran its tongues over the prey, tasting all the flavors. It lifted the prey, and, using its tongues, pulled the prey into its beak. It had to unhinge its jaws to accommodate the bulk of the prey. As it released the grip of its talons, raised the beak, and pulled with the tongue, the dragon drew the prey headfirst into itself.

The prey continued to struggle as it was swallowed, even within the alpha's belly until the digestive juices did their work. It could feel the rush of nutrients in its bloodstream as the prey dissolved, and it was glorious. It looked about, seeing its siblings feed as well, watched the prey continue to rush about the space, presumably looking for a way to escape. The alpha could feel its siblings' pleasure as they feasted. These prey were delicious and emotive, so satisfying and entertaining.

At one point, one of the prey, somehow, opened a barrier and ran into the corridor. One of the alpha's siblings flew after it, took it down, and consumed it in the corridor. The alpha assigned that sibling, for the good of the pack, to stay by that opening to prevent others from escaping. The sibling willingly set aside its hunger for the good of all, though it did manage to take two more prey that tried to escape. But, even this bounty was finite. As the remaining prey were consumed, the alpha searched through the adjoining chambers. It found a few smaller, presumably younger prey. Just as tasty and even more emotive!

So enjoyable…

When the last of the prey was consumed, when no more electromagnetic signatures were sensed, the insatiable alpha reached out with all of its senses and looked for more prey. It happened to look through one of the transparent panels in the barrier surrounding them. From there, it could see another set of prey moving in the distance through the foliage along the side of the structure. The alpha poked and prodded at the transparent panel until it found a latch. It fiddled with the latch until the panel popped out. A good, solid push forced the panel away, falling to the ground below.

The sound of its impact startled the prey. They looked around themselves, but they didn't look up. The alpha, increasingly excited, signaled its siblings. They gathered around the alpha by the opened panel. They could see the prey, now leaving the foliage to cross an open area. The alpha sensed their excitement; it felt their hunger.

It decided.

The alpha spread its wings and took off, hovering just outside as the others flew past. It followed them as they flew out, then hovered some 20 meters above the prey that still had not seen them. The alpha was cautious. It had seen some of these bolder prey out and about. The ones it had observed had fire and fought back. Something subtle about these prey reminded the alpha of those others. It did not want to lose any more siblings.

It decided.

Rather than attack the prey as a group, they would each pick one, then fall onto that specific one. There were more than enough of them; in fact, a few of the smaller siblings would have to hold back. Individually attacking the creatures would give the prey no opportunity to defend themselves. The alpha paused, waiting until it was clear which of its siblings would attack which of the prey. For itself, the alpha picked one it thought looked particularly tasty.

They tucked their wings and fell, hindlimbs bent to absorb the pending impact, talons spread to grasp their prey. Right before impact, the dragons spread their wings. The alpha enjoyed this very much. The prey remained oblivious as they silently fell. It was only when the alpha sank its talons into the back of its prey that the creature made a noise, a deep noise accompanying a significant exhalation when the wind was knocked out of it.

The alpha could hear and feel its siblings as they made contact, but it did not turn its attention from its prey. The prey collapsed beneath the impact, falling to the ground. Waves of shock and fear and anxiety wafted from the prey. The alpha folded its wings and raised its tails to counterbalance as it reared up on its hind legs. It opened its beak and extended its tongues stripping everything away. As it did so, as its tongues ran over the prey's flesh, the explosion of flavors thrilled the alpha. It raised the prey with its forelimbs, pulled the prey into its beak with its tongues, and initiated peristaltic contractions of its digestive organ to draw the prey deeper into itself. It felt its siblings do the same.

It was glorious. The alpha was nearly giddy with pleasure off the feeding compounded by the satisfaction of the successful hunt. It reared up again and raised its sensory cluster to scan the area. The alpha considered its options.

Suddenly, it felt a new kind of electromagnetic signature, neither sibling nor prey. It could sense the bewilderment of its siblings. The alpha lifted off, hovering a few meters above the ground; splaying its eyestalks, the alpha reached out with all its senses. The first impact was to its left wing. Suddenly, something sliced the webbing of its wing, shredding it, eliminating its ability to provide any lift.

It fell.

As it fell, the alpha felt more of the things hit. Dozens, then hundreds of blades sliced into its flesh. It felt the shock and pain and surprise of its siblings as it died.

▪ ▪ ▪

Pavel stepped back from his console, reached for the display he'd been viewing, and dragged it to the conference room table.

"Jane? Kim?"

They followed him as the display expanded. Jack and Keith walked over as well, followed by Yoshi as he returned to the room. Pavel hit replay as he said, "One of my squads was just attacked. Drones responded, but not quickly enough. They recorded this. I need to know whether we can help the survivors."

As they watched, dragons flew out through a broken window, hovered above a squad, then fell brutally upon them. The squad members didn't have a chance to raise their weapons.

Pavel winced, then continued. "This is the second dragon pack we've seen utilize this attack strategy. Seems like the larger dragons are hunting solo. It's the smaller ones, like these, that are hunting in packs." He saw Keith nod, then continued. "The self-organizing drones picked them up when they flew into the open, then destroyed them while they were attacking the squad." He choked up a bit as he said, "There are only three squad members left, and they're not moving..."

Kim turned to Jane. "I have some people with electromagnetic disruptors that could escort your people..."

Jane leaned in, reached over the display, and pulled her hands apart to zoom in. The bodies filled their point of view. She looked at their wounds and said, softly, "I'm sorry, there's no point." Pavel hung his head and walked away.

Yoshi sighed, pointed to one of the bodies, and said, "Peter's been with Pavel a long time. His wife's expecting their first child." After a moment, he turned to Kim and Jack and asked, "updates?"

"We have no idea how many dragons emerged under the residential dome," Kim said, "but we know the drones and squads have killed some 124 of them thus far. There were a few large solo hunters, I think thirty-one that we know we've killed. Haven't seen any in a while. The rest were smaller ones hunting in packs."

"Projections?"

"We think we're turning the tide. At this rate, we think we'll wipe them out within twenty-four hours."

"Any idea of our losses?"

"No, no idea yet."

Yoshi nodded, then asked, "what about the rest of the colony?"

Jack raised a finger. "We continue to see no further activity here under the primary dome. No signs of kraken or dragons. If you really think we're turning the tide under the residential dome, I'd sure like to reallocate some of the self-organizing drones to the rest of the colony. We have no idea what's going on over there."

"When's the last time you heard from any of our farmers?"

"It's been too long..."

DAY 21

SNATCHED FROM THE JAWS OF VICTORY . . .

Dawn broke. The baseline bank of artificial lighting along the eastern stem wall gradually brightened as the glow of the sun, obscured by storms, grew behind them. Chad had point; Amy brought up the rear. They walked along the roof of the residential building.

It had been a long night. Their last dragon skirmish was around 3am. Drones swarmed a particularly large dragon as it lumbered through a corridor. The drones nipped and sliced at it, unable to do serious damage in the tight space. Their squad arrived and hit it with flamethrowers, seeming to annoy it more than seriously hurt it. It backed away, slapping ineffectively against the drones with its talons. They came to a stairwell, and the thing backed up, climbing the stairs. Eventually, it emerged, stretching its wings and limbs once freed of the constraints of the building.

It didn't try to fly; Chad later commented he had the feeling it couldn't. Which their Eden friends suggested meant it might have been close to releasing spores. But, before anything else could happen, one of the larger drones appeared out of nowhere, apparently summoned by the smaller drones. It zoomed downward and pressed itself against the side of the dragon, right before the left wing. They heard a "whump," saw the dragon's body puff outward, then collapse.

They were too tired to cheer. Amy, Chad, and the rest continued to roam the structure, but they saw no further signs of the dragons. Then, as the dawn broke, the word came. Nobody had seen a dragon in over an hour. More tellingly, none of the drones had picked up an electromagnetic signature during that last hour, either.

Chad stopped and pointed at a set of tables, suggesting they take a break. He put down his flamethrower and walked to the roof's edge for a better view of the sunrise. Amy joined him.

"When's the last time you called in?"

He looked at his wrist computer and sighed. "It's been a while."

She nodded, put down her gear, and called in.

■ ■ ■

Yoshi convened the team, then turned to Pavel and asked him to repeat what he'd just said. Pavel nodded and addressed the room.

"We haven't seen a dragon or picked up an electromagnetic signal under the residential dome since 4am."

The team exchanged weary looks. "Are you saying it's over?" Jack asked.

Pavel shrugged. "It…appears to be over, at least under the residential dome. No signs of dragons, no signs of spores."

After a moment, Kim asked, "Then can we reallocate some drones to the agricultural domes?"

Yoshi, eyebrows raised, looked about the room; he saw weariness, but heard no objections.

"How do you want to do this?"

Kim cocked her head and rolled her shoulders. "Well, I'd start with a comprehensive view of the colony."

She pulled up the view of the residential dome they'd been referencing. She brought her hands together above the depiction to zoom out, then set the model to show the subterranean tunnels between the domes. The primary dome, enclosing the space elevator base station and colonial transportation hub, was on the equator. The residential dome was just west of the primary. Just west of that was the base, stem wall, and central tower of the second residential dome. Construction halted when the decision was made to accelerate the second settlement with its own orbiting platform and second space port. That decision preceded Kim's promotion to a level that would have given her a vote. She completely understood the logic behind that decision, but still hoped that, someday, construction would resume. South of the primary dome was the dome for the zoological and botanical park. North were the agricultural domes laid out in banks of five. East of the primary dome were the landing pads, and, south of the landing pads were the storage silos.

Those that could afford it took a landing craft between the landing pads and the orbiting platform rather than a space elevator gondola, but most of the landing pad traffic was freight. Materials shuttled to and from various projects around the planet, carrying freight too bulky or perishable for a gondola, livestock, for example. The silos were used to stage inbound and outbound freight. North of the landing pads were the industrial and infrastructure domes and fusion plants.

Kim tapped again to bring up the subterranean tunnels interconnecting the domes. The primary dome's base station served as the central hub for passenger traffic, connecting to the residential dome, zoological/botanical park dome, and the first bank of agricultural domes. The landing pad served as the central hub for freight traffic. Freight tunnels and shuttles were bigger and heavier duty than their passenger counterparts. They were designed for freight but also carried passengers, albeit in less comfort, and ran between the landing pads and the storage silos, industrial and infrastructure domes, agricultural domes, and the base station's fabrication center and gondola station.

She briefly considered highlighting tunnels blocked by shuttles, then decided not to make the display any busier. They were all on the same page.

"So..." Kim mused, "let's say we retrieve the drones from the residential dome. We want to release them in the agricultural domes. First question, how and where do we want them to disperse?"

Pavel tugged at his ear. "I'm not sure I care. We haven't heard from any of the farmers in a while. I'm hoping we'll find a lot of them alive, and huddled somewhere the kraken and dragons couldn't get at them. But I've no basis to suppose that's any more likely under one set of domes than the others."

Kim cocked her head. "Well, among the agricultural domes, I moved all the inter-dome shuttles to the middle of their tunnels; so, right now, they're all blocked. Whether that spared any of the agricultural domes is anyone's guess. But, I have to say, if there were any domes without kraken or dragons, I'd expect to have heard from those farmers by now."

Pavel shook his head. "I don't agree. But whether I agree or not is immaterial. We'll know soon enough."

Kim sighed. "Can't argue with that. Here's the thing, though. If I unblock the tunnels to let the drones fly freely between domes, we might also enable dragons to hunt domes they haven't been able to reach. But if I *don't* unblock them, then we'll have to carry through. That'll take a lot of time."

Yoshi volunteered. "Why don't you hand deliver them to some or all of the first bank of domes, the five directly connected to the primary. As each dome is either found to be clear or purged of dragons, you can open a tunnel to the next dome."

Kim, still staring at the image, nodded.

"What about your squads?" Yoshi asked.

"I think we should give them a break. I'm not ready to send them home, not yet. I don't think we'll need them for the agricultural domes, but I can't say for sure. Let's bring them back to the primary dome."

Yoshi turned to Kim and asked, "Can you spare a few people to set up that auditorium? Where we did the initial squad recruitment? I'm thinking some cots, some food?"

She smiled. "Of course!"

■ ■ ■

John walked in to find Elke staring out her window at the rains falling on the Garcia Mountains. He rested a hand on her shoulder, and she tilted her head to touch it with her cheek. She raised her head again to sip her coffee.

"How are you holding up?"

She smiled but did not turn. "Are you asking personally or professionally?"

He chuckled softly. "Professionally first. We'll get to the personal."

"Promise?"

"Yes, love."

She sighed, then turned to walk to her table. "They're making progress. They seem to have eliminated their kraken, now they're fighting dragons." She noticed John shudder at the memory of the dragons that raided his sick bay all those years ago.

She brought up a model of the Alexandria surface settlement. "They managed to clear the primary dome before those kraken reached sexual maturity. They also cleared their residential dome. Now, they're reallocating drones to the agricultural domes."

John nodded as he looked at the model. "I had a nice dinner with Dieter and Nikko while my significant other brooded here in her office. They caught me up on all of this. I must say, the self-organizing drones were a very clever idea. And blocking the tunnels by parking the shuttles midway? Brilliant!"

Elke swatted him with the back of her hand. "Brooding? You're one to talk. And, yes, they've improvised and come up with some excellent ideas. Now they need to triage their agricultural domes. The blocked tunnel may have spared some domes, but such isolation may have kept dragons in other domes at the cost of farmers and livestock." She grimaced, remembering equally difficult decisions she once had to make. She set all that aside, tilted her head, then decided to ask. "John, have you been keeping an eye on George?"

"In what respect? As a patient? He's recovering well, given his age. I have to admit, while I completely understand why you needed his help for Alexandria, I wish we could have spared him the stress. So, yeah, I'm keeping an eye on him." He smirked, then added, "you know what a mother hen I am. Why do you ask?"

She paused, not sure how much to say, then decided to go for it. "He's… different, since all this. And, I should tell you, I told him about the dual hub proposal."

John's eyes widened. "Elke, love, was that necessary? He's still healing from a heart attack. He's recovering from everything we've been through. And he's working his tail off trying to help that xenobiologist, what's his name…Keith, in Alexandria. This is something that could have waited!"

She stood and walked back to the window. "Maybe you're right. But I thought I needed to enlist his help. John, have you thought about the implications of this outbreak? They were already talking about trying to exterminate kraken and essentially terraform Eden. Then, we had the pineapple outbreak. Now, a deep space vessel crew's been lost, and God only knows how many Alexandria colonists…" She turned and added, "John, I'm afraid to open my messages! We don't have time to wait! He needs to start enlisting conservationists to fight this proposal!"

John hung his head a moment. "I get all that. And, as I've often said, you look after the people, I look after individuals. I'm worried about him. That's my job!"

"Well, John, my concern here is that his tone is changing. And, there's something about the words and phrases he's choosing. You know he's a religious man, right? But, he's starting to sound like one of these Edenists…"

John winced and drew a quick breath through his teeth. "Ah, ok. I understand. Well, I have to admit, since you told me about this dual hub thing, I've started thinking maybe these Edenists aren't crazy…" John stopped his musing when he saw the wildly concerned look in Elke's eyes. "Hey, I'll talk to George. Buy him a coffee, a decaffeinated coffee."

He hesitated, then approached her. "Love, there's something else we should touch on, especially given the terraforming talk. I've been communicating with the taxonomists. They're making progress cataloging Eden's biodiversity, but there's still so much work to. Anyway, it occurred to me that, if they focused their efforts to catalog the airborne spores of the trilateral superkingdom, we might be able to identify proteins shared by the spore casings. If we could do that, then, maybe, just maybe, I could come up with a vaccine."

She stared at him. "You mean we could walk the surface of Eden without environmental suits?"

He nodded. "Maybe. I think it's worth the effort. And, if we published the methodology, then, heaven forbid, if another deep space vessel were carrying an infected individual, or we inoculated departing deep space crews…well, we should talk about the potential."

▪ ▪ ▪

Yoshi and the senior staff gathered around the model. They decided to release three of the smaller, self-organizing drones in the five agricultural domes directly linked to the primary dome. They figured this would let them evaluate each dome's status and then take appropriate action. As they watched the model, Kim received updates from her people.

"Okay; drones were just released through the shuttle blocking tunnel to Dome 1. And now, Dome 3. Dome 4. Drones are en route to all five agricultural domes!" She listened, then, exhaling sharply, said, "Dome 4 is clear. No spores, no dragons. The farmers and their livestock are okay."

"Then *why* the *hell* haven't we heard from them?!" Pavel blustered.

She shrugged him off. "They weren't in their quarters. They gathered together in the central tower. They didn't receive any queries. Were you reaching out to them individually or their residences?"

Pavel grimaced but didn't answer her. Kim paused, then closed her eyes briefly and swore.

"Dome 3 is gone." It was her turn to grimace. "No signs of life, and there are spores. We'll keep those tunnels blocked, and the drones will stay there to start scrubbing." Kim paused again. "Dome 2 is also gone, but…Domes 1 and 5 are clear!"

She looked about the room. "We're unblocking the tunnels between the primary dome and those agricultural domes. My people will move their suspenser pallets through the tunnels and hubs, then to the shuttles blocking the next bank."

It took some time for the next collection of drones to make their way to the next set of tunnels.

"Dome 6 has dragons!" One of Kim's squad members exclaimed. "My team is releasing more drones to hunt the dragons under Dome 6! No word, yet, on survivors; this will take some time."

"Dome 8 is clear!"

"Dome 9 is gone. Spores."

"Dome 7 is clear!"

"Dome 10 has dragons!"

And so it went. As the anxious but hopeful group of officers watched, twenty agricultural domes in four banks of five became color coded on the model, signifying whether they were clear, infested with dragons, or contaminated with spores. Finally, as they reached the last bank, Kim, wearily, relayed status.

"Dome 19 is clear! Dome 18 is gone. Dome 16 is clear. Dome 17 is infested. Dome 20 is clear!"

They stared at the model. The weight of the color-coded model finally hit them. The promising colors of survival or chance. The dismay of the domes highlighted with death and loss. Yoshi broke the silence.

"Kim, please keep us appraised about the domes with dragons. We also need a plan to clear those domes of spores." He looked around the table. "People, is it over? I mean, once we clear these domes, is it, finally, over?"

Pavel turned to him with a rare half smile and nodded.

▪ ▪ ▪

"Elke!"

She turned from her conversation with John as George breathlessly entered the room. John's expression turned rapidly from happiness at seeing his friend to professional concern. He walked over and led George to a chair.

"Are you ok?"

"Hmm? Oh, yes. Yes, quite ok."

John pulled some instrumentation from his pocket and wordlessly went about a quick check-up as George turned to Elke.

"Governor, I think it's nearly over! I've just heard from Keith, and they think it's nearly over!"

Relief cascaded through Elke as she considered this news. She walked back to the projection above her conference room table and refreshed it. John straightened and put a hand on George's shoulder.

"You do, indeed, seem to be ok, George, at least physically. But let's try to take it easy, shall we?"

Elke bit her lip. She didn't want to stress George, but she needed to know what was happening.

"George, what am I seeing? Last I heard, the primary and residential domes were cleared. What about the agricultural domes?"

He stood, nodded at John, and walked over, pointing at domes as he spoke. "These agricultural domes are clear. These have been lost. There may or may not have survivors; drones haven't found any, but spores have been released. Kim's keeping them isolated. Some spores are flowing through the life support system, but they're being filtered out where they can. Drones with scrubbers are clearing the air under those domes. They'll need to clean the surfaces as well. Time will tell whether there are any leaks; but, if so, they'll be minor nuisances."

"What about the landing pads, zoological park, and so on?"

"Kim says they're next. They're isolated for now, but as soon as they finish clearing the agricultural domes, they'll check those. Keith is somewhat concerned about their zoological collection, of course, but there wasn't much other than possible vermin to keep kraken around by the landing pads or storage silos. They think that once they locked down all the tunnels, any kraken over there would have starved long before reaching breeding size."

▪ ▪ ▪

"Pavel? Jack?"

Kim stood by the settlement projection, watching update feeds from the drones. Pavel groaned as he rose from a cot; Jack walked over from the food counter. Jack reached her first and asked, "yes, Kim? What is it?"

She pointed to Agricultural Dome 6. "We have drones engaging dragons under a number of the domes, but they're reporting the dragon population seems to be dropping faster than they're killing them."

Pavel rubbed his eyes sleepily. "What do you mean?"

She blinked and pursed her lips. "I mean, they flew through the dome and did an initial count of the dragons, then started attacking them. They're now reporting how many dragons they've killed and how many dragons are left, and that's leaving quite a few unaccounted for. It's weird. We saw something similar under the residential dome, but I just thought that was because there were so many places to hide there. But, the agricultural domes don't have nearly as many structures. It's mostly open fields for crops and livestock."

Jack stared at her. "What are you suggesting?"

"I'm not suggesting anything! I'm just saying that we have a discrepancy. There are dragons we can't account for."

"Well, where would they go? You have the tunnels blocked, right?" He pointed at the freight tunnels between the agricultural domes and the tunnels between the agricultural domes and the landing pad. Kim nodded vigorously, then tapped her wrist computer to bring up the tunnel status by the landing pad. The indicator above each tunnel's intersection with the landing pad was red.

"See? The doors are closed."

Jack stared hard at the model projection, blanched, then yelled for Yoshi.

DAY 21

LAST STAND

Rafael was exhausted.

No doubt he'd rather be here than out there, fighting the kraken or dragons or whatever the hell they were fighting. But, damn, was he tired. They'd been running round the clock, making all kinds of stuff he'd never dealt with before. Sniffers. Flamethrowers. Weird ass drones. Now, they were outfitted with scrubbers. He wanted to go home and sleep in his own bed.

He stretched and resumed walking the aisles between the fabricators, checking status lights, talking to operators. He was, as always, doing his job. Then he heard something. Something like nothing he'd ever heard before. A rustling, or maybe a fluttering sound. He stepped towards the intersection of two aisles and looked around the fab center.

Then, from the freight elevator in the middle of the room, rose a cloud of nightmares.

■ ■ ■

Yoshi yawned as he walked over and then saw Jack's face.

"What is it?"

Jack spoke quickly but quietly, trying to control his rising panic. "We may have a very serious problem."

Yoshi froze, halfway through adjusting his hair. "What?"

Jack looked at Kim and pointed, but not accusatorily. "When Kim said she secured all the tunnels, I thought she'd moved shuttles to the middle of every tunnel. But, she just pointed to these indicator lights."

Yoshi blinked hard. "Oh, *shit*!"

Kim oscillated between them. "What??"

Jack held out a hand, pointed and exclaimed, "Those doors aren't there!"

"What do you mean 'they're not there'?!"

"I thought you knew. Happened before you took over. Those doors were never built. They were part of the original design, but we were behind schedule and, well, we never built those doors. And, I'm not aware of any surface settlement that's had a catastrophic decompression. Leaks, sure. But, nothing we can't handle. So, year after year, we prioritized other things and never got around to building those doors."

"But the fucking indicator lights say they're closed!!!"

Yoshi slowly walked past Jack and Kim towards the model, his face pale with exhaustion and dread. "Yes, the status indicators are there. The control circuits are there. But, the doors aren't."

Jack pointed as he spoke. "She says she can't account for all the dragons. The drones are reporting more reduction in the dragon population than they can confirm killed. I think they may have flown through the tunnels towards the landing pad!" Yoshi stared. Then Jack drew a deep breath and swore loudly.

"Kim, tell me you blocked the tunnel between the residential domes."

Kim stared at the model. "There *are* no tunnels between the residential domes! There's only the one dome!"

Jack grabbed his head in dismay; he muttered some swear words in disbelief. Yoshi kept staring at the display. Pavel leaned over and demanded Jack explain what he thought was happening.

"Don't you see??? We stopped construction of the second residential dome. We built the transportation hub, but, since we didn't finish the dome, we just sealed it off. The tunnels connecting the residential domes to each other and the second residential to the zoological dome are incomplete, but they're there! And, there's nothing blocking the landing pad. I think…I think the dragons escaping from the residential dome made their way as far as the zoo, and they're feasting. At some point, they'll run out of zoo animals, then fly through the freight tunnel to the landing pad. And the dragons escaping the agricultural domes are also flying through the freight tunnels to the landing pad."

They all turned to stare at the model, especially the freight tunnel between the landing pad and the primary dome's base station.

▪ ▪ ▪

"RUN!!!!"

Rafael screamed, and took off towards the exit. He didn't look back, though he heard screaming behind him as dragons fell on the operators. He made it into the corridor, ran some distance, then ducked into a utility room. He sealed the door behind him and called Kim.

▪ ▪ ▪

Kim ran from the display and yelled in the general direction of the petrified group "Stay there!" She paused at another station, hunched over to read something, then turned back, her eyes wide with terror.

"Dragons are pouring out of the freight elevator into the fab center!"

The room erupted with panic and chaos until Yoshi yelled, "HERE. *NOW*!" He turned to Pavel. "Where are your squads??"

Pavel tapped his wrist computer and red dots appeared in the display. Some were still in the residential dome transportation hub. Some were moving in a cluster, presumably aboard a shuttle traveling through the tunnel to the primary dome.

"They're on their way to the auditorium."

Yoshi turned to Jack and Kim. "Where else can the dragons emerge from that freight tunnel?"

Kim held her hands over the primary dome and spread them apart, zooming into the heart of the base station. "The fabrication center is at the lowest level of the base station. At the center is a freight elevator. The next level up is the transportation hub, and the next level after that is the space elevator gondola center."

She turned to Pavel. "I'm stopping that shuttle. We need to re-block the pathway to the residential dome and keep it clear. Your people aboard that shuttle need to exit the shuttle and proceed through the tunnel. Make sure they understand there are dragons ahead of them. The rest of your people need to start walking. They'll need to manually traverse that shuttle, using it like an airlock, to block the dragons."

She turned back to Yoshi. "I'm going to talk to my leads about the drones, see if there are any we can reclaim from the agricultural domes. We need to know where they're flying. We know they're going through the fabrication center. We should expect them in the transportation hub." She paused, then said, "it's the gondola center that worries me..."

Kim turned to Jack and asked, "Can we lock down the gondola center?"

He cocked his head. "Sure. Why?"

"What's in transit? Anything with passengers?"

Jack mumbled "I don't think so…" as he checked his wrist computer. "No. We suspended stalk operations when all this shit started. There were gondolas in transit, and some had passengers, but they've all arrived and been processed. There's nobody in that station. It's all locked down at the moment."

Kim clapped her hands. "Good! Excellent! Here's what we should do. We open the pathway from the transportation hub to the gondola station. The space elevator cables terminate there, and all the access in and out, for people and freight, is at that level. From there, the gondolas ride the cable up, through these central tubes and through the airlock at the top of the base station…"

"So, if we can drive the dragons into the gondola station…" Jack jumped in, anticipating her plan. "Then, we open that airlock…"

Kim looked at him and smiled, more out of belief than relief. "They'll be subjected to the planet's native atmosphere! I don't know whether they'll asphyxiate or fry first, and I don't care!"

Yoshi slammed his hand on the table. It was as good a plan as any. And they were out of time.

"What will that take?"

Kim paused to think. "We need to block every pathway we can." She tapped at her wrist computer, then said, "That locks down the fabrication center. I don't know how many dragons followed people out, but we'll deal with that later."

Jane looked at her with sudden concern. "What about the people still in the fabrication center? Won't they be trapped with the dragons???"

Kim nodded grimly. Pavel placed a hand on Jane's shoulder. "We have to focus on killing these things. And, to be blunt, I don't know that anyone who didn't already make it out is still alive."

Kim nodded again, then collected herself, then tapped at her wrist computer, saying, "And *this* locks down the transportation hub. They can fly out of that freight tunnel, they can roam the fab center and the transportation hub, but they have nowhere else to go except the gondola station."

Jack pointed at the projection. "If you open this portal at the top of the gondola station…"

"That opens the chimney at the center of the base station." Kim lifted her chin to acknowledge him. "That's a lot of air volume for the things."

"Before you do that, I have an idea. We have a couple of the electromagnetic disrupters in Keith's lab that we were using for testing. If we can tune those things to generate a magnetic field like a human, that ought to be effective bait. If we put fresh energy packs on them, they'll last a few hours."

Jack leaned in and pointed to the chimney. "There are maintenance access points at every level. If we position them here, inside the chimney, at our level, that could draw them up the chimney."

"Remember, there may already be dragons in there!" Pavel chimed in nervously.

"I know," Jack replied. "It's a chance we'll have to take. Do we have any weapons up here?"

Pavel nodded towards a corner of the room. "Not much. A few engineering samples over there. Should work, though."

Jack straightened his back and sighed. "I guess it's time we do our part." He pointed at the transportation hub. "That's locked down. Your squads are coming in through the tunnel. They can help drive dragons into the chimney. Maybe some can take the freight elevator down into the fab center and drive them out of there, too."

He turned back to Jane and added, "and rescue anyone that's still alive."

Keith scurried in the direction of his lab, emboldened. "Jane, let's go tune that disruptor!"

■ ■ ■

Yoshi walked through the corridor with Kim, Jack, and Pavel. "Y'know," he suddenly said, "it just occurred to me that it's possible any dragons that followed folks out of the fab center could be anywhere in the base station. Including up here!"

Pavel chuckled. He felt a lightness he could not explain. He'd aspired for this leadership position, worked his way up the organization for so many years. Yet, he could not remember feeling this way, not since he was a young man. Taking action, facing down his fears, not knowing what might be waiting around that next corner, those were the thrills, the sense of adventure that led him into security work in the first place. It wasn't that he had a death wish! But, riding a desk was not for him. He was going to need to think through what this all meant.

But not today. Today, his team needed him.

Jack took point, followed by Yoshi and Kim as they guided a suspensor pallet carrying two electromagnetic disruptors. One was active, tuned to detect dragons. Pavel brought up the rear. They initially considered deploying the disruptors on two different levels; but, they realized once the first was activated, the risk associated with deploying the second would be too high. Better to put them both out, then activate them and lock down the entry.

As they approached the portal, Kim stepped forward to open it, revealing a maintenance tube. She pondered a moment, then said, "hadn't thought about this. It's big enough to walk through, and big enough to push the pallet through, but not big enough to move past the pallet."

She pondered a bit more. "Okay, I'm going to pull it behind me, open the door, climb over them, turn 'em on, then run like hell out here to close this panel. Okay?"

Yoshi hesitated. "Wait…it's just, part of me thinks that's my job as your boss. But, honestly, given the space, I think you're the only one small enough to pull it off!"

She shrugged, then started pulling the pallet through the tube. Yoshi and Pavel posted on either side of the tunnel door. Jack stepped forward to push. She smiled at him nervously.

Once they reached the far end, she paused to address him. "You'd better head back. Once I open this door, I'm gonna be moving as fast as I can!"

Jack nodded and took a few steps back.

Kim braced herself. She mentally rehearsed her path, where she'd place her hands, where she'd place her feet, when she'd activate the devices. She turned, reached for the control, and pressed it. As the portal opened, she felt a rush of air. She hadn't expected a pressure differential; on reflection, it made sense, given that the chimney was part of an airlock sequence. She laughed at herself and turned to climb over the pallet as cruel talons reached in and pulled her into the chimney.

Jack stared in disbelief. He froze in fear and shock and panic as he heard her scream fade away. He pondered, briefly, whether there was any way to rescue her. Clearly, there wasn't.

Jack leapt forward and activated the devices. He briefly heard the flapping of leathery wings as he turned to run back through the tunnel. He'd taken his third step when he felt talons pierce his back. He felt his feet leave the floor as he fell forward; his chest hit the floor of the tunnel. Jack screamed, but his lungs were compromised, and it came out more a bloody gurgle. He reached forward with his right hand, trying to pull himself away as he felt the talons withdraw from his torso. He felt the weight of the dragon as one of its hind feet stepped on his thigh. The forepaws pressed against his shoulders and lifted him up as multiple, wet, muscular tongues ran over his body, tearing away his clothing.

Jack cried and pleaded for mercy as the tongues moved upward, as his body was lifted until his upper body was horizontal. His dangling limbs spasmed as he spastically struggled to free himself. He stared at the roof of the tunnel, continuing to cry out as he felt the tongues stretch from the top of his head over his face and ears; over his neck and shoulders, the tongues moved smoothly across his skin, then pressed against him. A thousand tiny knives pierced his skin.

Jack felt them pull him as he saw an open beak extend over his face. The beak clamped together as he was pulled deep into the muscular gullet of the creature.

Jack felt that muscle press against him, blocking his breath. Then the burning started. And then nothing.

■ ■ ■

Pavel and Yoshi stood by the door, staring down the tunnel. They heard Kim's screams, then they heard Jack call out her name. They saw Jack freeze, then lean over the pallet to activate the disruptors, then turn to run. They saw the dragon take him down. Yoshi started moving into the tunnel, but Pavel grabbed his arm and pulled him back.

"Did you see those wounds? He's gone! We need to close this door, NOW!!!"

■ ■ ■

Amy and Chad were surprised when their shuttle abruptly stopped midway between the residential and primary dome. They were even more surprised when they received revised instructions from Pavel. Amy shrugged, walked forward, and reached for the shuttle door manual control. She looked back at their squad, then met Chad's eyes. He nodded, so she turned to open the door.

The area surrounding the monorail was eerily quiet.

Amy climbed down and stepped to the right of the monorail, weapon ready, as the others climbed down. Chad sealed the door.

"What do you think? Do we wait for the rest?"

She chewed her lip. On one hand, there were people that needed help. On the other hand, stepping up to help them would be of little value if they were killed. And, it sounded like there were a lot of dragons up ahead.

"I think we wait here, ready to duck back into the shuttle. I'd guess the others got their instructions about the same time we did. They should already be walking. What's it take to walk a klick and a half? Fifteen, twenty minutes? Then we move forward together, flush these bastards towards that gondola chimney."

Chad nodded.

■ ■ ■

As Pavel and Yoshi returned to the conference room, Jane looked over. She read their facial expressions, then raised her hand to her mouth. "No…"

"What happened?" Keith asked incredulously.

Yoshi explained what they saw. The survivors were silent for a moment, then Yoshi said, "I know this hurts, but there are a lot of colonists counting on us to hold it together and finish this job. Jane, I need Pavel focusing on the task at hand. I need you to find people from Kim and Jack's teams that can step up in their place. Okay?" Still in shock, Jane nodded and stepped away.

Yoshi pointed at the model and asked, "can we tell whether it's working?"

Pavel shrugged, his head bowed. "We have limited visibility in there. We can see that there are dragons flying around, and we can guesstimate based on how often we're seeing them as to how many there must be, but that's about it. We really need some drones in there."

Jane walked back; Yoshi asked her who she'd found.

"I looked up who they usually delegate to when they're off duty. For Kim, it's a guy named Benny. For Jack, it's gonna be José. They're on their way up here. I, uh, didn't tell them why."

Yoshi sighed, then nodded. "Okay, well, while we wait, Pavel, where are your squads?"

Pavel leaned over and pointed to the shuttle. One set of dots was in the tunnel next to the shuttle. The rest were approaching the shuttle from behind.

"Maybe another five minutes. They can't all fit in the shuttle at once, especially given the bulk of the disruptors. They've been instructed to stage through, to not have both doors open at once. So, figure another five to ten minutes before they're all on the same side of the shuttle."

"How many?"

"Five squads total. Forty-seven people…given the losses."

"Well, we'll wait for them to all…"

A large man entered the conference room. Yoshi turned, then recognized him.

"Benny, thank you. We need your help. I need to know the status of the drones. Which are deployed in the agricultural domes, which can be reclaimed and redeployed here. Quickly, please."

Benny blinked, then asked after Kim.

■ ■ ■

Amy startled when the shuttle door opened, but she was relieved when the next two squads emerged. It took a bit of doing, to maneuver the suspensor pallets through the shuttle door; but, once they did, they were encouraged to have them hold the bulk of the disruptors.

As the squad members greeted each other, the leads moved away to confer. They quickly decided to move forward together with the disruptors activated to detect and deter the dragons. Once they reached the hub, they'd split into two groups. One to drive the dragons out of the hub, and the other to do the same in the fabrication center.

They moved forward cautiously. The disruptors detected dragons ahead of them, but none came within 100 meters of them. Once they reached the hub, the squads fanned out. They didn't see any dragons, but they could see their droppings here and there. They must have been roaming the hub until the disruptors came into range.

Amy turned to the other squad leaders. "I don't think we need to babysit these things. I say we set a few up across the hub with fresh packs and leave 'em running. We take the suspensor pallets with us. With only one disruptor per pallet, they'll be easier to maneuver."

Once the disruptors were set, they moved towards the freight elevator and pressed the button. After a bit, the doors opened. Amy, Chad, and the combined squads stepped through into the massive freight elevator car, pressed the fab center button, and felt the elevator descend.

The doors opened upon hell.

Dragons were hunting everywhere, some on the ground, others in the air. They could see bloody, shredded uniforms strewn across the floor. But, as the squads moved forward from the elevator car, weapons ready, disruptors activated, the dragons backed away. The squads moved to the corners of the fab center, positioned their disruptors, and stood by. As they watched, the dragons moved towards the freight elevator, then flew up the shaft. They pushed the suspensor pallets towards the shaft, then, when they reached it, looked up.

Chad called Pavel. "I can't say that every last one flew into the gondola chimney, but the disruptors are *not* registering dragons at all. If you don't have all of them, you should have most of them."

▪ ▪ ▪

The dragons were stressed. They didn't care for being confined. It wasn't a small space, but it was crowded with their kind. This crowding caused some of them to emit alarm pheromones, which increased the general level of agitation.

Above them, there were electromagnetic signatures of prey, but no actual prey to be found. Below them, there were maddening signals, signals they just could not tolerate. And, there was nowhere to light! They were flying, and they were starting to tire, but there wasn't anywhere to land except below them. Those signals were simply intolerable!

Suddenly, the dragons felt an electromagnetic flux below them and heard a smooth hum. They could see a flat surface growing out of the walls, coming together; the opening at the center shrunk, then closed completely. With great relief, they landed.

It was very, very crowded, and some of the larger, hungrier dragons contemplated which of the smaller ones they might consume. But before that could happen, they felt a similar flux above them. The dragons saw an opening grow above them as the flat surface blocking them shrank into the walls. And, as that opening grew, the temperature climbed sharply, faster and faster. And the air tasted terrible. They struggled to breathe.

Very quickly, the dragons realized they could not survive this much longer.

One by one, they released whatever spores their bodies had produced.

And then, one by one, they died.

PART 4
EPILOGUE

ALEXANDRIA

Yoshi stood by his conference room window, staring out past the dome interior at the roiling atmosphere beyond. He was exhausted, but so buzzed with stimulants he wouldn't be able to sleep for hours. Just as well, he had a great deal of work to do.

The colony was finally clear of the alien menace. He realized how melodramatic that sounded, and he chuckled softly to himself. Yoshi shook his head, stunned by the impropriety of his own reaction; he sighed before walking to his console to resume reviewing damage assessment reports. After Kim's brilliant inspiration worked, they'd gone about the process of finding and killing straggler dragons. Some released spores, but they were ready for that. They'd read Eden's reports; every time they killed a dragon, they had a second flamethrower ready to incinerate the spores as they were released.

Of course, they couldn't destroy them all. They had to assume some spores got away.

Jane worked with her counterpart, John at the Eden colony, on a recovery plan for Alexandria. Spore filters were added to every junction, and they fastidiously cleansed every surface to purge them of spores. A screening protocol was established for parasitic worm infections. Early detections would ease the applications of John's nanobot surgical procedures.

Yoshi also called for a review of "as built" discrepancies. They'd become complacent, letting things slide, year after year. But, there were reasons behind their blueprints, and letting things slide cost them dearly. No more!

They did manage to rescue a few people (but not nearly enough) from the fabrication center. The survivors were off duty personnel huddled in the storerooms that had been converted for cots. Thankfully, the dragons that made their way through the landing pad were smaller pack hunters. Larger dragons might have been able to apply their electromagnetic capabilities to push someone to open the portal. He'd seen reports that such things had happened under the residential dome.

Devilish things! Curse those dragons!

Yoshi once thought the kraken were pure evil. The dragon generation was even worse.

Evil, evil creatures. Evil incarnate!

He shuddered. He returned his attention to the reports. They had no idea how many colonists were gone. A lot, he knew. Benny, standing in for Kim, was working through an assessment, but it would take some time.

Poor Kim, and Jack…

They needed to work through memorial services, and they needed to work through recovery. He braced himself for the possibility of 10% losses, like Eden faced, which would be a mindboggling 20,000 souls. Losses would not only mean pain, but also breakage of the systems and processes that were the heart and soul of the colony! They'd need to rebalance their workforces, which would require a lot of retraining.

The assessment of their zoological park collection wasn't, relatively speaking, a priority, but it was straightforward. Every specimen was gone. It would take many years to rebuild their collection. The assessment of the damage to their agricultural capacity was anything but straightforward, and early reports were grim. They'd lost a lot of farmers. Livestock were devastated. What little livestock survived would need to be screened and potentially treated for parasitic infections. John had little guidance to offer regarding the efficacy of his nanobots for livestock. Livestock that couldn't be cleared would have to be destroyed. They simply couldn't take the chance that worms might infect consumers, despite cooking and human digestive processes.

Crops were mostly intact. They would need to be inspected for spores and cleansed where necessary. It looked like the kraken destroyed a lot of their bees. Eden reported the bees apparently irritated the kraken, and they weren't kidding. Yoshi had no idea, yet, of the potential impact to their crop futures given the loss of so many hives.

The domes themselves were contaminated to varying degrees, a number of them severely. Clearing them could take weeks, or maybe months, which led to some potentially horrific math. What if they didn't have enough food to sustain the survivors? Sure, they had food stores, but that was finite. It'd been *years* since they had to rely on algal synthetics. At this point, those fabricators mostly made supplemental livestock feed. They could reallocate, and potentially ramp up, fabrication capacity for human consumption, but that would take time. Though not as long as rebuilding agricultural capacity. And, there were inbound migrants aboard deep space vessels. No way to reach them; no way to turn them back. That influx should be balanced by outbound migrants, and they could try to encourage more emigration.

But, ultimately, there was no escaping a real possibility of insufficient food quantities, even with rationing. Especially given the dire preliminary estimates he reviewed. Yoshi shook his head again as he considered the potential implications.

Given these projections, if the kraken and dragons haven't killed enough of my people, we may face starvation!!! He stood abruptly and returned to the window. *No, dammit! There'd be enough.*

They might have to deal with nasty rationing protocols, but there'd be enough! He shook his head. Nothing more he could do now, not until they had more information.

Meanwhile, he owed Elke a status update. The Eden team had been invaluable to their struggle; and he assumed Eden would be invaluable to their recovery. For this, he was grateful.

But, Yoshi couldn't deny his smoldering resentment. He'd found in the records a warning from Governor Lubandi to then Governor Kelty, his predecessor. He knew that, after the departure of the *Lucky Strike*, every person aboard every deep space vessel leaving Eden was screened for spores. But, that just wasn't enough. He'd dug into everything he could find, and done more than his share of soul searching, and he couldn't, for the life of him, come up with a single thing Elke should have done and didn't, or didn't do and should have, or should have done differently. And still, thousands of the colonists who put their faith in Yoshi Gulati's leadership were dead, with considerably more hardship ahead of them. That had to be *somebody's* fault, wasn't it?

And it was possible that the *Eureka* was carrying kraken to yet another colony…

He also needed to send a status update to his leadership. He knew he needed to be professional, dispassionate. But that would be challenging. Some way must be found to bind the terrors of Eden *to* Eden. Yoshi slammed his fists against the glass with surprising strength, causing the pane to shake. He hissed through gritted teeth.

"This *cannot* be allowed to happen anywhere else!"

EDEN

George stood in his lab, staring absentmindedly at a collection of Eden marine life. The aquarium was a 4 meter wide, 2 meter deep cylinder. At the center was a rock outcropping, encrusted with life, that had been transplanted from the sea east of the Garcia Mountains. A riot of shapes and colors moved slowly through and over the outcropping, with a variety of living forms of such diverse shapes and colors swimming about it, the whole aquarium staggered the imagination.

The lab, a massive room on the top level of the base station, was filled with containers of various types holding living and preserved specimens of Eden's staggering biodiversity. But, the aquarium was not a research project. It wasn't purely ornamental, either, whatever people might say. For George, it was both therapeutically calming and a constant reminder of the wonders he strove to understand and preserve. Having just reviewed Keith's latest updates, the calming quality was particularly important. Alexandria seemed to have finally cleared their infestation. The kraken and dragons engendered mixed emotions. Horror was the primary feeling, of course. He often pondered whether anything in human history had ever been as terrifying. Of course it was hard for modern humans to judge. Seeing large predators like lions or bears or crocodiles in a video or as part of a zoological collection did not, *could* not, do justice to the terror they once inspired for our ancestors. But, have any of these creatures ever been as terrifying as kraken or dragons?

At the same time, how could one not admire them? Exquisitely adapted to capitalize on the chaos created by Eden's tectonic dynamism, more effective than any life form ever encountered. Indeed, it was the frequency and intensity of Eden's natural disasters that created such an ecological niche! They were incredibly robust and adaptable, flexible and clever, aggressive and ferocious.

Every life form balances adaptation to their ecological niche with variability that provides the evolutionary fodder enabling them to respond to challenges and opportunities when their niche changes. When an environment endures an upheaval, a fire, a floor, a tsunami, a quake, any of the natural disasters, that environment is reset. It undergoes a succession, a gradual return to maturity. On earth, after a fire, grasses become shrubs become trees. Along the way, a variety of species adapted to exploit temporary opportunities to thrive and reproduce before restoration completes. Eden's constant churn facilitated the evolution of a variety of species whose specialization is generalization, the capacity to adapt to nearly any conditions, even one that is, for them, completely alien.

And, among those specializations (not uncommon among succession pioneers) were abilities related to establishing a foothold. On Earth, there were a variety of creatures that could reproduce asexually, if that's what it took to gain said foothold. It only took one pioneer to reach the new environment. Hell, aboard their deep space vessels and across the colonies of known space, they were plagued by parthenogenetic mutations of such vermin as roaches and crickets. Despite their best efforts, despite vacuum purges and traps and cats and everything else, it seemed there'd always be at least one of the damn things that would survive.

One was all it took to reinfest.

Here, through the exposure of one man to one spore, kraken nearly established a foothold on Alexandria. *One* spore. The implications were staggering. And, George knew, sooner or later, these implications would be recognized and understood by those that did not share his passion for stewardship.

Some of this was not new. Every human settlement was a fulcrum balancing tension between preservation and development. But, nowhere else in known space was there such a fulcrum, such a need for preservation, such a drive for development.

He understood what David had found. He wasn't a planetologist, but a competent evolutionary biologist, and ecologists had to have a working knowledge of the planetological context of the environments they studied. And so, he regularly reviewed David's publications. He understood the tidal forces of Eden's four moons were driving tectonic activity that not only drove the upheavals and successions he studied, but also drove the formation of ore bodies David and Cecilia wanted so desperately to develop. And, he understood they'd found evidence of the fusion of neutron stars, millennia ago, that apparently seeded the Eden system with precious metals, rare earths, and stable superheavy elements.

The importance, and value, of this discovery was not lost on him. Humanity, throughout known space, depended on technology that, in turn, depended on precious metals and rare earths. And, the potential utility and value of superheavy elements? Substances historically only producible through intensely expensive artificial processes? The prospects were staggering.

So, this is what he faced. On one hand, greed. Well, not just greed, that wasn't fair. Certainly, there were those seeking to enrich themselves, but such enrichment was only possible because these materials were so desperately needed. He'd recently read, though not fully understood, a published physics paper theorizing the possibility that properties of such superheavy elements might, someday, facilitate development of faster than light travel. Imagine the exploration potential! Imagine finding more Edens! But at what cost? So, not "greed" per se, but "development." Yes, that was a fairer term.

Development versus stewardship. A tension as old as time. Tenuously balanced on that fulcrum But, now, that awful "dual hub" proposal threatened to place a thumb on that scale. And, now, fear of interstellar infestations would add another thumb.

What to do?

George stood and began pacing, oblivious to specimens and experiments in progress that would normally fascinate him, struggling to sort through what to do next.

▪ ▪ ▪

Elke stood by her window. She'd just received Alexandria's latest status update. The dragons were gone, but the recovery was just beginning. She had some idea of what that would be like, though she also knew Yoshi was facing this on a scale that her mind struggled to comprehend. Elke shook her head and sighed.

They'd do all that they could. She could not have asked more of her team than what they'd accomplished, especially on the heels of the pineapple incident. Everyone was exhausted, everyone needed time to recover. She also knew there'd be a call for some sort of accountability. Part of her believed she'd done all that could reasonably be expected of her; part of her would never be satisfied with what she had and hadn't done. She also knew that, for the most part, this wouldn't be seen as a human failure, but rather a natural hazard. And that was far worse.

Humanity wasn't used to dealing with alien predators. In all of known space, only Earth and Eden had complex ecosystems. On Earth, very few creatures remained outside of zoological parks and reserves that could threaten human life. Before Eden, there'd been generations, centuries, since the last time a wild creature killed a person. Eden was a wilderness, and wilderness had predators. And, most humans across known space might find that notion intellectually interesting but were not prepared for the actuality. It was one thing to receive news that a colonist had been lost to a predator on Eden. But, to hear an Eden predator devastated *another* colony?

Elke shook her head again. The forces pressing for unfettered development of Eden would use this. They'd use this, they'd use the dual hub concept, and they'd push *hard* for terraforming Eden. They might push hard enough to push her aside. Not that Elke cared about her career at this point. She was ready to retire. But she wanted to retire *here*, and she didn't want to retire to a terraformed Eden. So, she'd stay. And she'd use every tool at her disposal to remain in the Governor's seat, and leverage that seat to protect Eden.

It wasn't going to be easy, and David Fessler, bless his heart, just made her job a whole lot harder.

She had some time to catch up on her messages during Alexandria's crisis. Before the pineapple incident blew up, David sent her a proposal. A proposal for a marine research station. She loved the idea. And she knew, and she knew that David knew, that George would love it, too. But, she knew David, and his wife, Cecilia. So, she pored through his recent scholarly publications.

Sure enough, he'd found rare earths and stable superheavy elements in relatively high concentrations in polymetallic nodules near Eden's sea-floor hydrothermal vents. Something about their concentration in the hot fluids coming out of the vents and then forming nodules around seed objects. She didn't really understand all of it, but she knew that David's motivation for proposing this deep-sea research station wasn't purely academic, which was ok. As Eden's Chief Planetologist, discovering such potential wealth was his job, just as it was Cecilia's job, as their Chief Mining Engineer, to propose methodologies to safely, responsibly, and profitably harvest such wealth.

She'd summoned George to her conference room an hour before. He hadn't heard about the marine research proposal, but he wasn't surprised, either. She smiled ruefully. She was familiar with the interesting relationship dynamics between George and David and Cecilia. There was mutual respect, sure, and, though they often tried to hide it, real enjoyment of each other's company. But there was also suspicion, questioning of motives.

Her job, of course, was to balance the wealth potential discovered by David with that discovered by George. They'd already identified so many valuable organic chemicals! And, even after fifteen years, they'd barely scratched the surface of Eden's incredible biodiversity, which meant it was possible, even probable, there were discoveries to be made that rivaled anything found by David.

She knew she was supposed to approach her job pragmatically. She could defend every time she'd prioritized the preservation of Eden's biodiversity as a pragmatic recognition of the perishability of that biodiversity. Once a species is gone, it's gone, along with any potential value it may have offered. Who could guess what wonderful opportunities had been squandered, undiscovered, during Earth's Great Extinction of the 21st century? But, she knew it was more than pragmatism. Elke knew she had a bias towards preservation, a bias she had to leaven cautiously, lest her favoritism be called out. Hers was the opportunity, the responsibility, for stewardship of the first alien wilderness in known space. She had to set the precedents that would guide her successors, and that required reigning in her preservationist inclinations lest she be removed prematurely from her position.

She turned her head from the window to briefly glance at the table, remembering how frail George looked when he entered the room. Yes, they'd all been through hell, but George was still recovering from that heart attack. After he'd taken a seat, she'd explained David's proposal. They'd reviewed a draft projection of the proposed facility, and George suggested numerous improvements to support his department's research needs.

"George," she'd said, "I need to broach a potentially sensitive subject with you."

"Of course! Please, speak freely!"

"George, I'm going to need your help here. These rare earths are desperately needed. Earth's reserves are gone, with demand exceeding what the rest of the system can provide. Very few colonies are net exporters. If David's right, if these nodules are as rich as he thinks they are, there will be tremendous pressure to harvest them. I've no doubt Cecilia will do her best to develop responsible extraction methods, but…"

He nodded along. "I know, Elke. We don't know, we *can't* know whether Eden's sea floor ecosystems are as sensitive as Earth's, but I can't offer any reason to believe they aren't! And we know what kind of damage this kind of deep sea mining did on Earth."

"Exactly. I'll do everything I can to support your enhancement proposals of David's plans to support impact research, but I can't block development. And, if he's right about the super-heavy elements…"

George raised a hand and said, "Then the question won't be whether to harvest, it'll be the least bad way to do it. And 'least bad' will be a relative thing." He shook his head, then continued, "trust me, I understand! I'm well aware of the demand for rare earths. I know how important these materials are to the equipment I employ in our research, and to the quality of life I enjoy! And, superheavy elements? Stable, harvestable superheavy elements? My God, the potential!"

They held each other's gaze for a moment.

"George, there's something else. I'm concerned about the time you're spending with the Edenists."

He stared back, raised his eyebrows briefly, then nodded. "They have a point, you know. I do believe we're being tested." He'd gestured towards the window and the Garcia Mountains. "We've been given this opportunity, and our stewardship is being tested. Tested, *again*, after a dismal failure with respect to Earth."

He turned to the window and continued, "It's easy, well, *relatively* easy, to balance development and preservation in the best of times. But these are not the best of times. And, I swear, it's almost as if God is intentionally ratcheting up the temptation with these discoveries! But, if we're to prove ourselves truly worthy of this opportunity, then we need to demonstrate responsible stewardship despite the challenges and temptations!"

It was her turn to nod. "I agree with you, George. That's why I shared this dual hub proposal with you. And, it's why I'm sharing David's proposal with you, and why I've shared my concerns about a backlash after Alexandria's kraken infestation." She reached out to hold his wrist, then added, "And it's why I need your credibility to be impeccable! I need, *we* need, for you to distance yourself from the Edenists."

"Governor, my faith is my business!"

She withdrew her hand and meshed her fingers as she paused.

"Yes, George, your faith is your business. My commitment to the stewardship, the preservation, of Eden's biodiversity is rooted in my firm, secular belief in our responsibility to future generations, not in religious faith. But you know, or at least I hope you know, how much I respect your faith. But, George, the Edenists are extremists. They're not, at least not in anything I've heard or read, pragmatic. We cannot deny the need to harvest Eden's resources. That WILL happen, one way or another! We cannot stop that, we can only manage it. Manage it to ensure such harvesting preserves Eden."

He stood, then walked to the window. "What gives us the right?" Before she could answer, he turned from the window and raised a hand at her. "That was a rhetorical question. But I can't deny my empathy for their positions!"

"George, I will NOT ignore the hypocrisy of those that would PIP their way here only to decry our presence on this planet. I will NOT."

She collected herself, then continued, "I will do everything I can do to preserve Eden. But, I won't deny reality, and I will *not* condone extremism. I will *not* condone vandalism or sabotage, and I will *not* tolerate terrorism!"

He stared at her unfazed. "Is that what you think?"

"George, I've heard the same speeches you have. And, while I haven't heard calls for vandalism or violence yet, I think that's the path they're on."

She paused, then said, "Let's set that aside, George. Let's presume, for the moment at least, that the rhetoric won't progress to that point. But, George, then what? They talk and talk and talk, but they don't *do* anything! Can you name one action they've taken? Even proposed?"

She leaned back and shook her head. "Talk is cheap! It's cathartic, sure. Hearing each other blather about the importance of preservation and then going to the pub *feels* good, George, but it *doesn't accomplish a goddamn thing*!"

Elke took a deep breath, then added, "My fear, George, is that if they *do* take action, it'll be counterproductive. Terrorism. Sabotage. Something that appeals to them in the moment, but is ultimately counterproductive. Something that invalidates their arguments, and actually shifts the debate in the favor of development! Can't you see that?"

She stood and walked past him to point out the window. "Eden needs action, George. *Productive* action! Action pragmatically rooted in the reality of our situation. I'm asking you to distance yourself from the Edenists for your own sake, and to partner with me to preserve all that!"

She turned to him and added, "I need your credibility to be untarnished, because I need your voice to be heard and trusted across known space. I need you to share, with everyone, what's at stake here, to make the case that everything here, even the ugly parts, the kraken and the pineapples and God only knows what other challenges we'll meet, must be preserved!"

George lowered his eyes, then smiled at Elke. She'd gotten through. At least, Elke *thought* she had. They didn't have a plan of action, not yet. But, she believed George would work with her rather than waste time with extremists. She knew that raising awareness through George's popular audience would be an important part of the equation. If she and George could just find a way to recruit David's voice to speak for the preservation of Eden.

The key, she felt, would be David and Cecilia.

NOTES AND ACKNOWLEDGMENTS

I'm grateful to my family for putting up with me locked away in my office for so many hours of writing and rewriting (and, of course, re-re-writing). In particular, I want to thank my wife for encouraging me along the way. Diana, you are my muse!

The folks at Paperclip Publishing are a joy to work with. Thank you especially to Brooke for her faith in me, and Abigail for her exceptional (and patient) editing wizardry (she doesn't just catch my typos and run-on sentences; she's made numerous suggestions that made it a better book!). Thanks to Hannah for her diligence and artistry, and Shaun for his brilliant, beautiful depictions. Thank you also to my fellow authors. I enjoy your company, camaraderie, creativity, and partnership more than I can say.

I am most especially grateful for the growing community of *Deep Space Exodus* readers. Nothing can compare to the feeling of seeing someone's taken the time to post a review, or the feeling of talking with a happy reader at an event who's read *Kraken* or *Tides* and wants *Return*! And a special thank you to the fans (Kay was the first!) that asked for this book (you know who you are).

Thank you to Erik K. who, in another conversation, happened to say just the right thing for me to realize what the book was missing and made all the difference.

Hours were spent working through the underlying science for *Return from Eden*. A big part of the fun (for me and hopefully for readers) is exploring Earthly biological examples that inform what might be possible on an alien world. I do love writing these books and I hope you've enjoyed reading the story as much as I've enjoyed writing it.

Thank you!

ABOUT THE AUTHOR

George Moakley started his career studying biology, with dreams of doing fieldwork and ecosystem modeling. To make ends meet, he took a data entry position with a precious metals company. Throughout his long career in the tech industry, George gained an inside perspective on various scientific and business practices, including designing Edge Intelligence solutions and conducting strategic planning workshops. He owns several patents, and holds prestigious advisory positions with ASU's School of Business and UCI's Customer Experience Program.

A fan of science fiction since childhood, George remains fascinated by the hard sciences. He feeds his passion for nature through photography while traveling the world; he also regularly partakes in hiking and scuba diving. Such travels have brought great joy, but also great concern about the fragility of our ecosystems. He has witnessed devastation caused by climate change, invasive species, overfishing, and other environmental issues. George is a long time member of the Nature Conservancy; a meeting with a conservancy representative in 2019 inspired a strategic thought experiment, regarding how the twenty-first century is likely to play out. The results were sobering, but provided the real-life inspiration behind Kraken of Eden.

Today, George makes his home in sunny Arizona with his inspiration, Diana. Between them, they have seven kids. When he's not locked away writing, he loves to visit his children and grandchildren, and can often be spotting driving back and forth from various sports activities.

www.ingramcontent.com/pod-product-compliance
Lightning Source LLC
LaVergne TN
LVHW011525200926
848484LV00001B/153
* 9 7 9 8 9 9 8 7 8 2 1 7 6 *